# Farming, Food and Nature

Livestock production and its use of finite resources is devastating biodiversity and pushing wildlife to the brink of extinction. This powerful book examines the massive global impact caused by intensive livestock production and then explores solutions, ranging from moving to agroecological farming to reducing consumption of animal products, including examples of best practice and innovation, both on land and within the investment and food industries.

Leading international contributors spell out the problems in terms of planetary limits, climate change, resources, the massive use of cereals and soy for animal feed, and the direct impact of industrial farming on the welfare of farmed animals. They call for an urgent move to a flourishing food system for the sake of animals, the planet and us. Some offer examples of global good practice in farming or the power of the investment community to drive change, and others highlight food business innovation and exciting developments in protein diversification. Providing a highly accessible overview of key issues, this book creates a timely resource for all concerned about the environmental, social and ethical issues facing food, farming and nature. It will be an invaluable resource and provide inspiration for students, professionals, non-governmental organisations (NGOs) and the general reader.

**Joyce D'Silva** is Ambassador Emeritus for Compassion in World Farming, the leading charity advancing the welfare of farm animals worldwide. She is co-editor of *The Meat Crisis* (second edition 2017).

**Carol McKenna** is Special Advisor to the Chief Executive of Compassion in World Farming and organised the Extinction and Livestock Conference on which this book is based.

## Other books in the Earthscan Food and Agriculture Series

**Farmers' Cooperatives and Sustainable Food Systems in Europe**
*Raquel Ajates Gonzalez*

**The New Peasantries**
Rural Development in Times of Globalization (Second Edition)
*Jan Douwe van der Ploeg*

**Sustainable Intensification of Agriculture**
Greening the World's Food Economy
*Jules N. Pretty and Zareen Pervez Bharucha*

**Contested Sustainability Discourses in the Agrifood System**
*Edited by Douglas H. Constance, Jason Konefal and Maki Hatanaka*

**The Financialization of Agri-Food Systems**
Contested Transformations
*Edited by Hilde Bjørkhaug, André Magnan and Geoffrey Lawrence*

**Redesigning the Global Seed Commons**
Law and Policy for Agrobiodiversity and Food Security
*Christine Frison*

**Organic Food and Farming in China**
Top-down and Bottom-up Ecological Initiatives
*Steffanie Scott, Zhenzhong Si, Theresa Schumilas and Aijuan Chen*

**Farming, Food and Nature**
Respecting Animals, People and the Environment
*Edited by Joyce D'Silva and Carol McKenna*

For further details please visit the series page on the Routledge website:
http://www.routledge.com/books/series/ECEFA/

# Farming, Food and Nature

## Respecting Animals, People and the Environment

**Edited by Joyce D'Silva and Carol McKenna**

First published 2018
by Routledge
2 Park Square, Milton Park, Abingdon, Oxon OX14 4RN

and by Routledge
711 Third Avenue, New York, NY 10017

*Routledge is an imprint of the Taylor & Francis Group, an informa business*



*British Library Cataloguing-in-Publication Data*
A catalogue record for this book is available from the British Library

*Library of Congress Cataloging-in-Publication Data*
Names: D'Silva, Joyce, editor. | McKenna, Carol, editor.
Title: Farming, food and nature: respecting animals, people and the environment / edited by Joyce D'Silva and Carol McKenna.
Description: Milton Park, Abingdon, Oxon; New York, NY: Routledge, 2018. | Series: Earthscan food and agriculture | Includes bibliographical references and index.
Identifiers: LCCN 2018028203 | ISBN 9781138541412 (hbk) | ISBN 9781138541443 (pbk) | ISBN 9781351011013 (ebk)
Subjects: LCSH: Livestock—Social aspects. | Livestock—Climatic factor. | Livestock—Moral and ethical aspects.
Classification: LCC SF140.S62 F37 2018 | DDC 636—dc23
LC record available at https://lccn.loc.gov/2018028203

ISBN: 978-1-138-54141-2 (hbk)
ISBN: 978-1-138-54144-3 (pbk)
ISBN: 978-1-351-01101-3 (ebk)

Typeset in Bembo
by codeMantra

# Contents

# Contributors

## Editors

**Joyce D'Silva** has an MA from Dublin University, Trinity College, and has been awarded honorary doctorates from the Universities of Winchester and Keele. She has taught in schools in India and the UK. She wrote the UK's second vegan cookbook, *Healthy Eating for the New Age* Wildwood House, 1980), which ran to four editions.

In 1985 Joyce began a long career at Compassion in World Farming (Compassion), becoming Chief Executive in 1991, a post she held for 14 years. Since April 2016 she has been working as Ambassador Emeritus for Compassion on a consultancy basis. She played a key role in achieving the UK ban on sow stalls in the 1990s and in getting recognition of animal sentience enshrined in the European Union (EU) Treaties. She has given evidence to the UK and New Zealand governments, the European Commission and the UK Government's Advisory Committee on Climate Change.

Joyce addressed (and organised, with the Royal Society for the Prevention of Cruelty to Animals, RSPCA) the first conference on pig welfare in Beijing in 2005; initiated Compassion's 2005 Conference "From Darwin to Dawkins: the Science and Implications of Animal Sentience"; and, in 2017, helped organise the conference "Extinction and Livestock: moving to a flourishing food system for wildlife, farm animals and us". Joyce co-edited (with John Webster) *The Meat Crisis: Developing more sustainable and ethical production and consumption*, Earthscan, 2017. She has received several awards for her animal welfare work and was the joint recipient of the RSPCA Lord Erskine Award in recognition of a "very important contribution in the field of animal welfare" in 2004.

Joyce is a Patron of the Animal Interfaith Alliance and a Trustee of the UK branch of Help in Suffering, a charity carrying out veterinary work in Jaipur, India. She is on the Advisory Council of CreatureKind.

Joyce has authored chapters for several books and reports, including *Welfare Challenges* ed. Dr Andy Butterworth, CABI, publication pending 2018; *The Impact of Meat Consumption on Health and Environmental Sustainability*, eds

Raphaely, T. and Marinova, D., IGI Global, 2015; the *UNCTAD Trade and Environment Review 2013*, United Nations Conference on Trade and Development (UNCTAD); and *Global Food Insecurity: Rethinking Agricultural and Rural Development Paradigm and Policy*, eds Behnassi, M. and Draggan, S., Springer, 2011.

**Carol McKenna** is a successful campaigner and coalition builder with over 30 years of experience of a broad range of animal protection issues. She is currently Special Advisor to the Chief Executive of Compassion in World Farming, leading projects and work strands aimed at achieving systemic change towards regenerative agriculture. She led the organisation of the 2017 Extinction and Livestock conference hosted by Compassion in World Farming, World Wide Fund (WWF) and others, which inspired this book (www.extinctionconference.org).

Carol is also a Trustee of the Eating Better alliance (www.eating-better.org) and a founding member and Trustee of the Wild Animal Welfare Committee, a charity which provides independent advice and evidence about the welfare of free-living wild animals in the UK (www.wawcommittee.org). Other recent roles include Campaign Director of Compassion from September 2015 to October 2016 and Director of Programmes at WSPA (World Society for the Protection of Animals), now World Animal Protection, from 2009 to 2011. Whilst at WSPA Carol initiated the Animal Protection Index project (www.api.worldanimalprotection.org).

For many years, Carol was a consultant to a number of international organisations. Projects included coordinating and co-chairing the Beyond Calf Exports Stakeholders Forum 2006–2013 and co-ordinating from 1999 to 2004 the Campaign to Protect Hunted Animals, the coalition that achieved the ban on hunting with dogs in England and Wales.

Carol is the author, joint author and/or editor of many reports, including *The modern solution to the exports of calves: working in black and white*, The Beyond Calf Exports Stakeholder Forum, 2013; and *Bidding for Extinction – snapshot survey of ivory trade on eBay*, IFAW, 2007.

Most recently, Carol has written a chapter, "The impact of legislation and industry standards on farm animal welfare", in *The Meat Crisis: Developing more sustainable and ethical production and consumption*, edited by Joyce D'Silva and John Webster, Earthscan, 2017.

Carol studied at University of Aston in Birmingham and has a BSc in Modern Languages.

## Authors

**Aysha Akhtar**, MD, MPH, is a neurologist and public health specialist with the Office of Counterterrorism and Emerging Threats of the Food and Drug Administration, and a Fellow of The Oxford Centre for Animal Ethics. She also serves as Lieutenant Commander (LT CDR) in the US

Public Health Services Commissioned Corps. She is the author of the book *Animals and Public Health. Why treating animals better is critical to human welfare.* Visit her website at www.ayshaakhtar.com

**Jasmijn de Boo** is International Director of ProVeg International and former CEO of SAFE (Save Animals from Exploitation) in New Zealand and The Vegan Society in the UK. She has worked in the animal protection movement and education sector for over 18 years.

**Peter Borchert** is an independent commentator, publisher, conservationist and lifetime protagonist for the value of Africa's wild places and cultural heritage. His media and publishing career spans some 45 years. As the founder and publisher of *Africa Geographic* magazine, he won high regard for his vision and probing editorials on and about Africa. Today, his abiding commitment to Africa, its wildlife and its people finds expression through his website, Untold Africa, and a new venture – The Wildlife Review Foundation – the purpose of which is the production of an ongoing series of interactive, multimedia websites that explore the conservation issues of our time. The first project, the *Rhino Review*, is already underway and is scheduled for publication in the latter half of 2018.

**Natasha Boyland** is Fish Policy and Research Manager for Compassion in World Farming, where she is responsible for managing the development of technical resources and lobbying strategy for improving fish welfare. She also contributes to the Aquaculture Advisory Council, writing research reports for discussion by major EU aquaculture stakeholders. Natasha's research background is in the field of animal behaviour. Following her BSc in Biological Sciences at the University of Reading, she gained an MSc and PhD at the University of Exeter, where her research focussed on animal communication networks and the effects of social behaviour on farm animal welfare. She continues to collaborate with the Centre for Research in Animal Behaviour at Exeter as an Honorary Research Fellow.

**Phil Brooke** is a biologist and is Research and Education Manager for Compassion in World Farming. His role has evolved into helping to develop the animal welfare knowledge base in Compassion's Research and Education Department, providing technical support for colleagues and appearing in the media. He has a reputation within the organisation for his knowledge of farm animal welfare issues, built up over a lifetime of studying the subject, visiting farms, listening to experts and producing a range of briefings and educational materials. He has also supported Alison Mood, his partner, in the development of her Fishcount website, which addresses the welfare of both wild and farmed fish. Phil is Vice-Chair of the Fish Working Group of the EU's Aquaculture Advisory Council.

**Donald M. Broom** is Emeritus Professor of Animal Welfare at Cambridge University, Department of Veterinary Medicine. His research concerns the scientific assessment of animal welfare, cognitive abilities of animals, ethics of animal usage and sustainable farming. He was Chairman or Vice Chairman of EU Scientific Committees on Animal Welfare 1990–2012. He has published 360 refereed papers and 11 books. His recent books are *Sentience and Animal Welfare* (CABI, 2014), *Domestic Animal Behaviour and Welfare, 5th edn* (CABI, 2015), a 75-page report for the European Parliament *Animal Welfare in the European Union* (2017) and *Tourism and Animal Welfare* (CABI, 2018).

**Chris Clark** is an enthusiastic but amateur environmentalist. With his wife he has established an 180-hectare hill farming enterprise in the Yorkshire Dales. Native breeds have been introduced, livestock numbers have been lowered and trees have been planted, with a subsequent increase in profitability and biodiversity. Formerly he was a tenant farmer, where he established a successful free-range meat venture. He then went on to develop a business management consultancy primarily dedicated to the agriculture and allied food industries. He is a member of the Yorkshire Dales National Park Authority.

**Glyn Davies** is Senior Adviser in WWF Malaysia, working on conservation as well as the sustainable production of palm oil and timber. He was Conservation Director for WWF-UK (2007–2017), leading work to safeguard the natural world so that we have a planet where humans and nature thrive. Prior to WWF-UK Glyn served as Director of Conservation Programmes at the Zoological Society of London. Throughout his career he has worked at local, national and international levels, looking at environment and development, forest and wildlife conservation, and biodiversity policy.

**Karl Falkenberg** has enjoyed a long career at the Commission of the EU. He was an EU trade negotiator during the Uruguay Round and subsequently negotiated the first World Trade Organization (WTO) Services Protocols in Financial Services, Telecommunications and Maritime Transport. From 2003 to 2008, he was in charge of all bilateral trade relations as Deputy Director General for Trade. From 2009 to 2015 he was Director General for Environment, covering all domestic and international environmental issues, including Climate Change. From July 2015 to July 2017 he served as Senior Adviser to Commission President Jean Claude Juncker on Sustainability. He drafted a Report on Sustainability and spent the final year as Senior Fellow at Oxford University. Karl left on retirement on 1 July 2017 and now works as a lecturer on trade, environment and sustainability. He is a trained economist and journalist.

**Bruce Friedrich** is co-founder and Executive Director of The Good Food Institute (GFI), a non-profit organisation that promotes innovative alternatives to industrial animal agriculture. He has penned op-eds for the *Wall Street Journal*, *USA Today*, the *Los Angeles Times*, the *Chicago Tribune* and many other American publications. Bruce is a popular speaker on college campuses and has presented repeatedly at most of America's top universities, including Harvard, Yale, Princeton, Stanford and the Massachussets Institute of Technology (MIT). He has co-authored two books, contributed chapters to seven books and authored seven law review articles. Bruce graduated magna cum laude from Georgetown Law and Phi Beta Kappa from Grinnell College. He also holds degrees from Johns Hopkins University and the London School of Economics.

**Leah Garces** was until end September 2018 the USA Executive Director at Compassion in World Farming (www.ciwf.com). She is an animal advocate who partners with some of the largest food companies in the world to improve their animal welfare policies and practices. Her work has been featured in many national media outlets, including the *New York Times*, the *Washington Post*, Buzzfeed, Vice Magazine and the *Chicago Tribune*, amongst others. She is a contributing author to Huffington Post and Food Safety News, and a member of the *Forbes* Nonprofit Council. In 2016 alone, the Compassion US team partnered with global food companies to make commitments that will improve the lives of over 300 million farmed animals annually. Leah serves as the Chair of the Global Animal Partnership, one of only three animal welfare certification programmes in the US. In October 2018 she became President of Mercy for Animals. She is also the mom of three incredible kids.

**Dave Goulson** is Professor of Biology at the University of Sussex. He studies the ecology and conservation of bumblebees, and he has published over 270 scientific papers on bees and other insects. In 2006 he founded the Bumblebee Conservation Trust, a membership-based charity in the UK which now has 12,000 members and has helped to create over 1,000 hectares of flower-rich habitat for bees. He is also the author of the amazing books *A Sting in the Tale*, *A Buzz in the Meadow* and *Bee Quest*.

**Elena Hemler** is a Program Coordinator in the Department of Nutrition at the Harvard T.H. Chan School of Public Health. Her main interests include the prevention and treatment of malnutrition and obesity in underserved populations, sparked by her previous work as a teacher in Tanzania. Elena holds a BS in Biopsychology and Community Health from Tufts University. She completed a thesis with the Tufts Friedman School of Nutrition, studying the efficacy of an educational intervention targeting the caretakers of malnourished children in the USAID Title II Supplementary Feeding Program in Malawi. She has served as a health coach for overweight patients at Boston Children's Hospital and a New Sector Alliance fellow, working to improve early detection of developmental delays in low-income children. Elena was previously a project manager at Epic, an electronic health records company, where she implemented healthcare technology solutions in hospitals around the world and worked to improve nutrition clinical documentation.

**Hans R. Herren** is President of the Millennium Institute (2005-), which develops scenario models and approaches to assist countries and development partners in the design, implementation and reporting on the Sustainable Development Goals (SDGs) with inclusive multi-stakeholder processes. An agricultural scientist, he was awarded the World Food Prize in 1995 for his work on biological pest control at the International Institute of Tropical Agriculture (1979–1994). In Nairobi he ran the International Centre of Insect Physiology and Ecology, with emphasis on the integration of human, animal, plant and environmental health (1994–2005). He is a passionate believer in sustainable, ecological agriculture; co-chaired the International Assessment of Agricultural Knowledge, Science and Technology (IAASTD); and is founder of the Biovision Foundation based in Zürich.

**Patrick Holden** is founder and Chief Executive of the Sustainable Food Trust, an organisation working internationally to accelerate the transition towards more sustainable food systems. Between 1995 and 2010 he was Director of the Soil Association, during which time he pioneered the development of UK and international organic standards, policy incentives for organic production and the organic market.

His policy advocacy is underpinned by his practical experience in agriculture on his 100-hectare holding, now the longest established organic dairy farm in Wales, where he produces a raw milk cheddar-style cheese from his 80 native Ayrshire cows.

Patrick is a frequent broadcaster and speaker, and was awarded Commander of the British Empire (CBE) for services to organic farming in 2005 and an Ashoka Fellowship in 2016.

**Frank Hu** is Chair of Department of Nutrition, Fredrick J. Stare Professor of Nutrition and Epidemiology at Harvard T.H. Chan School of Public Health and Professor of Medicine, Harvard Medical School and Channing Division of Network Medicine, Brigham and Women's Hospital. Dr Hu received his MD from Tongji Medical College in China and a PhD in Epidemiology from University of Illinois at Chicago. His research is focussed on nutritional and lifestyle epidemiology, prevention of obesity and cardiometabolic diseases, gene-environment interactions and nutritional metabolomics. He has published more than 1,000 papers (H-index 215) and a textbook on Obesity Epidemiology (Oxford University Press, 2008). In 2010, Dr Hu received the American Diabetes Association Kelly West Award for Outstanding Achievement in Epidemiology. He has served on the Institute of Medicine (IOM) Committee on Preventing the Global Epidemic of Cardiovascular Disease; the AHA/ACC Obesity Guideline Expert Panel; and the 2015 Dietary Guidelines Advisory Committee, USDA/HHS. Dr Hu was elected to the US National Academy of Medicine in 2015.

**Tony Juniper**, CBE, is the Executive Director for Advocacy and Campaigns at WWF-UK. He began his career as an ornithologist, working with Birdlife International. In 1990 he began a long period working at Friends of the Earth (FOE), initially leading the campaign for the tropical rainforests and in 2003 becoming Executive Director. Among FOE's many achievements during Juniper's time there was the success of the Big Ask campaign, which called for a new law on climate change (leading to the 2008 Climate Change Act – the first legislation of its kind in the world). Tony was the first recipient of the Charles and Miriam Rothschild conservation medal (2009) and was awarded honorary Doctor of Science degrees from the Universities of Bristol and Plymouth (2013). In 2017 he was appointed CBE for services to conservation. On joining WWF in January 2018 Tony remained a Fellow with the University of Cambridge Institute for Sustainability Leadership and President of the Wildlife Trusts. He is the author of many books, including the multi-award-winning 2013 bestseller *What has Nature ever done for us?* His latest book, *Rainforest – dispatches from Earth's most vital frontlines*, was published in March 2018.

**Andrew Knight** is Professor of Animal Welfare and Ethics, and Founding Director of the Centre for Animal Welfare at the University of Winchester; an EBVS European and RCVS Veterinary Specialist in Animal Welfare Science, Ethics and Law; an American Veterinary Specialist in Animal Welfare; a Fellow of the Royal College of Veterinary Surgeons; and a Senior Fellow of the UK Higher Education Academy.

**Philip Lymbery** is a naturalist, author and Chief Executive of the leading international farm animal welfare organisation, Compassion in World Farming, and Visiting Professor at the University of Winchester. His latest book, *Dead Zone: Where the Wild Things Were* (Bloomsbury, 2017), exposes how cheap meat is a key factor in the demise of some of the world's most endangered species. For 25 years Philip has worked extensively on animal welfare issues, wildlife and the environment. He has played leading roles in major animal welfare reforms, including Europe-wide bans on veal crates for calves and barren battery cages for laying hens. Philip is a recognised thought leader and has a reputation as one of industrial farming's fiercest critics. Described as one of the food industry's most influential people, he has spearheaded Compassion's engagement work with over 800 food companies worldwide, leading to real improvements in the lives of over 1 billion farm animals every year.

**Karen F. Mancera** completed a PhD in Veterinary Science, focussed on Animal Behaviour and Welfare at the University of Queensland, Australia, under the supervision of Prof. Clive Phillips. Her MSc in the National Autonomous University of Mexico (UNAM) focussed on the effects of different percentages of tree coverage on cattle behaviour and welfare in the tropics. Currently, Karen works as a postdoctoral fellow in UNAM, integrating socio-economic, environmental and animal welfare indicators for the evaluation of sustainability in extensive cattle systems in Latin America.

**Janet Maro** was born in 1987 in Kilimanjaro. She is a graduate of Sokoine University of Agriculture (SUA) with a BSc in Agricultural Economics and Agribusiness and is finalising a MSc in Project Management. She has had a strong passion for sustainable agriculture, which crystallised during her work for Bustani ya Tushikamane (ByT), which works with farmers on organic farming. In 2010, she launched Sustainable Agriculture Tanzania (SAT) to make it possible to act on the national level. SAT is a grass-roots organisation where ecological agriculture is practised. It always works by involving farmers in the planning stage and other stakeholders, like universities, companies and governmental extension officers, in its activities. This holistic approach has helped to establish an Innovation Platform, where dissemination, research, application and networking build the main pillars. SAT has worked with 101 farmer groups (2,700 farmers) in 70 villages in Morogoro, where food production and livestock keeping in an environmentally friendly and sustainable way is practised.

**Joseph Okori** joined the International Fund for Animal Welfare (IFAW) in 2017 to lead their landscape conservation programme and their Southern Africa Regional office based in Cape Town, where he directs their work to save elephants, rhinos, lions and other African wildlife through the protection of critical landscapes, mitigation of human-wildlife conflict and ecologically sustainable community development projects. Previously Dr Okori led the global rhino conservation programme and spearheaded Africa-Asia trans-continental partnerships on rhino and elephant conservation for WWF out of their South Africa office. He also served as the WWF focal point for the South African Development Community (SADC) in strategy development. Over his 19 years in conservation Dr Okori has led innovative ranger and research projects, focussing on DNA forensics as a tool for wildlife crime enforcement as well as building strategic partnerships and advocating for long-term policy protections for wildlife in national and international fora.

**Martin Palmer** is Secretary-General of the Alliance of Religions and Conservation (ARC), an international organisation founded by Prince Philip in 1995. ARC inspires religious groups to develop programmes on the environment and ethical food. Martin has written English editions of Chinese philosophical classics such as Zhuangzi as well as books on world faiths. He co-chairs an ARC environmental programme with the United Nations Development Programme (UNDP). He regularly contributes to TV and radio programmes on faith matters.

**Raj Patel** is a writer, activist and academic, holding a research professorship at the University of Texas, and Senior Research Associate at Rhodes University (UHURU). He has worked for the World Bank and WTO, and protested against them around the world. In 2016 he was recognised with a James Beard Foundation Leadership Award. Raj co-taught the 2014

Edible Education class at University of California, Berkeley with Michael Pollan. Raj has published widely in economics, philosophy, politics and public health journals, and the press. His first book was *Stuffed and Starved: The Hidden Battle for the World Food System*; his next, *The Value of Nothing*, was a New York Times bestseller; and his most recent is *A History of the World in Seven Cheap Things*, co-authored with Jason W. Moore. He is currently working on a groundbreaking documentary project about the global food system with award-winning director Steve James.

**Michael Pellman Rowland** serves as a Wealth Advisor at Alpenrose Wealth Management, guiding clients on sustainable investment. He was named to the 40 under 40 list by *Investment News* and was recently included in the *Forbes* Top Next Generation Advisors list. He is a frequent guest lecturer and has been featured in *The Wall Street Journal, Financial Times* and *Institutional Investor.* In addition, Michael is a contributor at *Forbes*, covering the future of food as it relates to sustainability. He recently organised a future of food event with Jeremy Coller, discussing the investment risks of factory farming.

**Jonathon Porritt** is co-founder of Forum for the Future, the UK's leading sustainable development charity. He was formerly Director of Friends of the Earth, Co-Chair of the Green Party and Chairman of the UK Sustainable Development Commission, and is the Chancellor of Keele University. Recent books are *Capitalism As If The World Matters* (2007) and *The World We Made* (2013), which seeks to inspire people about the prospects of a sustainable world in 2050. Jonathon received a CBE in January 2000 for services to environmental protection.

**Katherine Richardson** is Professor in Biological Oceanography at the University of Copenhagen and Leader of the Sustainability Science Centre (www.sustainability.ku.dk). She has recently been appointed by the UN as General Secretary to the 15-man expert panel to draft the global sustainability report that will inform the 2019 UN General Assembly. She is one of the coordinating authors on the "planetary boundaries" framework aimed at defining a safe operating space for human perturbation of the Earth System. She is a principle investigator in the Center for Macroecology, Evolution and Climate (www.macroecology.ku.dk). She chaired the Danish Government's "Climate Commission", which provided a plan for how the country could become independent of fossil fuels and formed the background for the parliamentary goal of removing fossil fuels from the energy and transport systems by 2050. She currently sits on the Danish Council on Climate Change and is active both as a member in and/or chairperson for a number of national and international research committee/advisory boards. She has published over 100 peer-reviewed scientific papers and book chapters.

**Carl Safina** is the first Endowed Professor for Nature and Humanity at Stony Brook University, where he co-chairs the Alan Alda Center for

Communicating Science and runs the not-for-profit Safina Center. His writing about the living world has won a MacArthur "genius" prize, Pew and Guggenheim Fellowships, and several book awards and medals. He hosted the Public Broadcasting Service (PBS) series *Saving the Ocean*. He is widely published in leading magazines, and his books include the classic *Song for the Blue Ocean*. His latest book is *Beyond Words: What Animals Think and Feel*.

**Jimmy Smith** is Director General of the International Livestock Research Institute (ILRI), a position he assumed on 1 October 2011. Before joining ILRI, he worked for the World Bank, where he led the Bank's Global Livestock Portfolio. Previously, he held senior positions at the Canadian International Development Agency (CIDA). Still earlier in his career, Smith worked at ILRI and its predecessor, the International Livestock Centre for Africa (ILCA), where he led an association of 10 consultative group for international agriculture research (CGIAR) centres working at the crop-livestock interface. Smith has also held senior positions in the Caribbean Agricultural Research and Development Institute (CARDI). He is a graduate of the University of Illinois, where he completed a PhD in animal sciences. He is widely published, with more than 100 publications.

**Peter Stevenson** is a qualified lawyer educated at Trinity College Cambridge. He is the Chief Policy Advisor of Compassion in World Farming. In 2004 he was the joint recipient of the RSPCA Lord Erskine Award in recognition of a "very important contribution in the field of animal welfare". Peter is lead author of the study by the UN Food and Agriculture Organisation (FAO) reviewing animal welfare legislation in the beef, pork and poultry industries. He has written well-received reports on the economics of livestock production and on the detrimental impact of industrial farming on the resources on which our future ability to feed ourselves depends.

**Pat Thompson** is Senior Land Use Policy Officer (uplands) for the Royal Society for the Protection of Birds (RSPB) based in Newcastle. He holds a BSc (Hons) and a PhD, and has worked for the RSPB since 1995. In his current role, Pat leads on upland policy, working with researchers, policymakers and site-based colleagues across the UK. Prior to his current role, he held a number of roles in the RSPB, including Conservation Officer (Caithness & Sutherland) and Regional Public Affairs Manager (North England). Pat is a passionate advocate for the UK's mountains, moors, hills and valleys, and the benefits they provide for wider society.

**Jean-François Timmers** is WWF Global Leader on Soy, based in Brazil. A biologist with specialisation in Economics and an MSc in Ecology, Conservation and Sustainable Development, he has led conservation, development, advocacy and research projects and programmes in Brazil since 1992, including the "Man and the Biosphere – MAB" Program of

UNESCO. He also coordinated an innovative programme on food security and co-management of parkland with indigenous communities as well as the creation of 13 new federal protected areas in the Atlantic Forest, leading to the protection of more than 315,000 hectares of highly threatened last remnants of rare ecosystems. He joined WWF-Brazil in 2012 to set up its Public Policies programme, dealing with a wide range of policy, legislative and conservation issues, including the Soy Moratorium in the Amazon. In 2015, he joined WWF International as Global Leader on Soy and since then he also led the initial structuring of WWF's new global action on food, coordinating actions from WWF and partners to protect the world's largest deforestation/conversion front from the expansion of soy and cattle ranching in the Brazilian Cerrado as well as other threatened South American ecosystems.

**Rosie Wardle** is Programme Director at the Jeremy Coller Foundation (JCF), a strategic grant-making organisation based in London, established in 2002 by entrepreneur Jeremy Coller, founder and Chief Investment Officer of Coller Capital, the leading global player in the private equity secondaries market. In her role as Programme Director, Rosie leads the Animal Welfare and Human Health programmes, which address the global sustainability consequences of factory farming and human health threats, including antibiotic resistance. In December 2015, JCF launched the Farm Animal Investment Risk & Return (FAIRR) Initiative, which is working with major investors globally to put factory farming on the Environmental, Social and Governance (ESG) agenda. FAIRR is currently backed by investors with combined assets of over $4 trillion, and Rosie leads the investor engagements with global food companies. She also acts as an Advisor for Jeremy Coller's personal investments in the food technology space. Recent investments include Memphis Meat, Impossible Foods and Perfect Day. Rosie holds a BA (Hons) and MA from the University of Oxford, and an MA from the Courtauld Institute in London.

**John Webster** graduated as a vet from Cambridge in 1963 and is now Professor Emeritus at the University of Bristol. On appointment to the Chair of Animal Husbandry at Bristol (UK) in 1977 he established the unit for the study of animal welfare and behaviour, which is now over 60 strong. He was a founding member of the UK Farm Animal Welfare Council and first propounded the "Five Freedoms" that have gained international recognition as standards for defining the elements of good welfare in domestic animals. He is a former President of the UK Nutrition Society and British Society of Animal Science. His most notable publications include two books on animal welfare: *A Cool Eye towards Eden* and *Limping towards Eden*. His most recent book, *Animal Husbandry Regained: the place of farm animals in sustainable agriculture*, extends the principle of unsentimental compassion beyond that for farm animals to embrace all concerns for the life on the land.

**Duncan Williamson** is a global food system expert and has been working in the field of sustainable systems for 20 years. He is the Food Policy Manger for WWF-UK, heading their food work. He devised and delivers the ongoing Livewell project, which demonstrates that a healthy diet can be sustainable. He is leading the WWF Network's area of collective action on sustainable diets, focussing on Asia and South America. He is on the steering group for their work on the post-2015 agenda. He is coordinating WWF's work on sustainable food security, including the recently published report *From Individual to Collective action: exploring the business cases for addressing sustainable food security*. He also advises companies on food sustainability issues. He has run projects in Greece, in Thailand and on the Galapagos Islands. He has degrees in Sustainable Environmental Management and Philosophy. He is a Trustee of Eating Better, on the Advisory Board for FCRN and Food Bytes and the UNSCP ten-year programme on food, and on the steering groups of Protein 2040 and Peas Please.

**Krzysztof Wotjas** is Head of Fish Policy at Compassion in World Farming. He is currently leading a Europe-wide project aimed at improving the welfare of fish in the EU and the UK. He is also an Assistant Professor at the Wroclaw University of Environmental and Life Sciences. He has conducted various research projects and published scientific papers on animal welfare with a special focus on fish cognition and welfare. Previously he was a consultant for Eurogroup for Animals, a Brussels-based organisation lobbying in the European Commission and Parliament for animal welfare. He is also working with Otwarte Klatki, a Polish-based NGO, where his scientific report contributed to a parliamentary initiative for an outright ban on fur farming in Poland.

**Dominic Wormell** is Head of the Mammal Department at Durrell Wildlife Conservation Trust, an organisation dedicated to species conservation. He has worked for Durrell for over 25 years and is an internationally renowned expert on the captive management and conservation of marmosets and tamarins. He has contributed to several in situ recovery and reintroduction programmes for these small monkeys in their native South America. He is devoted to building up the skills needed by conservationists to care for these sensitive primates in their own countries. He was involved in the first-ever reintroduction to the wild of the black lion tamarin and has been instrumental in generating support for the restoration of this species' habitat.

# Foreword

During my lifetime there has been a huge increase in the number of people around the world eating more and more meat, and this has led to the growth of industrial animal farming. Today, many people are aware of the tremendous cruelty to the billions of animals in these "factory farms", and there is also growing understanding about the effect of overusing the antibiotics they are fed routinely – just to try to keep them alive. However fewer people understand the devastating impact of these factory farms on the environment and wildlife so I was delighted when Compassion in World Farming and WWF organised a conference to discuss this desperately important issue.

Huge areas of habitat have already been devastated for growing soya and grain to feed these billions of imprisoned farm animals. Massive amounts of fossil fuel are burned to take grain to animals, animals to slaughter and meat to table. Much water is wasted in transforming vegetable to animal protein and the methane gas produced during digestion is contributing to the greenhouse gases that have led to climate change. In addition forests are being gradually desertified by the grazing and browsing of cattle and goats; and the bushmeat trade is affecting millions of wild animals.

This important book is the result of that important conference. The various issues that were discussed are presented here by authors who have in depth understanding of the topics they write about. It is a book that provides information that will help people understand how desperately we need to reform industrial animal farming – and our own eating habits. There are already farming methods, which respect and nourish the earth and its ecological diversity, which care for wildlife, support hard working farmers, and improve human health. Now is the time for us to adopt these.

Finally, I hope that this book, by pointing out the damage we're doing to our planet through intensive animal farming and the worldwide increase in the consumption of meat and dairy products, will serve to inspire readers to play their part in helping to achieve the urgent change that is needed. Each one of us can play a role in this, not only governments, corporations, investors and farmers, but we the consumers. If millions and eventually billions of us make ethical choices as to what we eat and how it was produced we shall help to bring about the change that is needed if we care about the planet and future generations. Time is short. We must take action now.

Jane Goodall, PhD, DBE
Founder - the Jane Goodall Institute &
UN Messenger of Peace
www.janegoodall.org
www.rootsandshoots.org

# 1 Introduction

*Joyce D'Silva and Carol McKenna*

There's a sense of urgency in the air. The evidence for rapid climate change is growing; biodiversity and wildlife are obviously in trouble; the very soils on which we all depend for food are losing their vitality; water is becoming scarce and polluted; and, scandalously, poverty and hunger are still with us. Our planet itself is in turmoil.

It was to address these fundamental issues that Compassion in World Farming (Compassion) decided to bring the best minds and hearts together at a major conference in London in October 2017. Partnering with World Wildlife Fund (WWF)*, we called the conference "Extinction and Livestock: moving to a flourishing food system for wildlife, farm animals and us". Over 500 policymakers, academics, students and interested citizens spent two days listening, conversing and making decisions.

This book brings all that innovative thinking together – and expands it with contributions from experts who couldn't be at the conference or learned about it too late! The chapters form a timely resource of detailed reviews, policy explorations, snapshots, case studies and personal reflections. We hope this diversity reflects the wealth of knowledge and activity on this vital topic and the high level of engagement among numerous groups, from researchers and students to non-governmental organisations (NGOs), activists, decision makers and the concerned general public. Indeed all of us are concerned about the huge environmental, social and ethical issues posed by our current food systems.

You might ask what have livestock got to do with extinction or soils or water – or even hunger? Hopefully, by the time you've read this book, you will understand the links and see why current industrial agricultural practices are threatening the very planetary systems upon which we all depend.

This is why we begin with a look at those planetary boundaries, the limits beyond which we should not venture. We ask one of the world's leading experts, Katherine Richardson, to explain this to us and to show how industrial livestock farming is a major factor in pushing these boundaries to the edge – and maybe, beyond. It's a deeply worrying scenario.

Compassion's CEO, Philip Lymbery, takes the lessons he has learned from researching his two highly regarded books, *Farmageddon* and *Dead*

*Zone – where the wild things were*, and shows how keeping farm animals in the indoor confines of the factory farm has direct impacts on the lives of many wild animals, from the Sumatran elephant to the Brazilian jaguar. Tony Juniper, one of Britain's best-known environmentalists, explains, with passion, why this type of farming has led to massive ecosystem damage, is causing extinction of species and must be reformed.

Dave Goulson, best friend to bees, argues that the continual spraying of crops with pesticides is endangering the very pollinators on whom agriculture depends – surely a disaster in the making. From his experiences of tamarin and marmoset rescues and reintroductions, Dominic Wormell challenges conservation organisations on the meat issue, calling on them to tackle the habitat devastation caused by beef production and soy, most of which is grown to feed animals kept in intensive farms. His call for change, especially in South America, is echoed strongly by Jean-François Timmers, whose work for WWF in Brazil brings him into constant contact with this issue.

Fish expert Krzysztof Wotjas explains how the very oceans are being depleted of life, not just to feed humanity but to be crushed into feed for animals in industrial farms, some of them also fish species! Taking small fish from the ocean to feed larger fish kept in cages is surely folly of the first order! Fish farming certainly raises a host of environmental and animal welfare issues.

Academic and passionate campaigner Raj Patel savages the cheap food paradigm, which subjects animals to misery and the people who work in the industry to low wages and often horrendous conditions. Leading global policymaker Jimmy Smith makes a strong case for supporting small-scale livestock farmers, whose families and livelihoods depend on their animals.

Award-winning author Carl Safina takes an innovative look at some of our animal brothers and sisters. "Who are you?" he asks, making the case for seeing creatures as living their own lives, with their own relationships, rather than as predator, pest or potential food. Joyce D'Silva uses her experience of many years at Compassion in World Farming to outline the appalling ways in which we breed, feed, mutilate and confine billions of farm animals globally in order to fill our shelves and plates with cheap food.

Don Broom, one of the world's leading experts on the welfare of farm animals, makes a strong case for treating these sentient beings with respect and care, if we want to continue eating their products.

John Webster, author of some of the most readable books on farm animal well-being, makes a strong case for keeping fewer animals in more natural conditions, letting nature's ruminants, like cows and sheep, graze the land, preserving and possibly enhancing environmental diversity.

Several people with huge experience give us practical examples of making kinder, agroecological or organic farming work in very different settings: Janet Maro in Tanzania; Karen Mancera in Mexico; Joseph Okori and Peter Borchert in Malawi; and Patrick Holden, Pat Thompson and Chris Clark in the UK. These real-life case studies are vital in demonstrating the variety of agroecological approaches available.

Looking to the global picture of food policy – what is working and what is not – we have economist and lawyer Peter Stevenson, who shows the inefficiency of feeding grains and soy to intensively kept farmed animals. The economics just don't add up! This issue has huge implications for the Sustainable Development Goals (SDGs). Some goals may simply be unachievable if we don't change the way we farm. Two leading experts in this area, Hans Herren and Karl Falkenberg, delve deeper and both conclude that reform must happen – and happen fast! Jonathon Porritt adds a rousing condemnation of what he so aptly terms "the productivist fantasy".

Martin Palmer, who works with the major faiths to achieve environmental reform, relates a wonderful tale of how individual action can achieve big results. He urges us not to ignore the power of the faiths to bring about change.

Although the more novel issue of farming insects for human food or animal feed was not addressed at the conference, we feel it is becoming such a hot issue that it should be included in the book. Biologist Phil Brooke does a sterling job of explaining all the associated issues, from economic and protein efficiency to the question of insect sentience.

Carol McKenna creates an enthusiastic overview of action being taken by businesses, civil society, cities and others to persuade people to eat less meat and more plants or "meat" alternatives. Practical initiatives that are driving change right now are suggested as models for others to adopt, adapt or replicate.

Dietary sustainability and healthy eating are well covered by Harvard global public health specialist Frank Hu and his colleague Elena Helmer. They are convinced by the medical evidence on heart disease, obesity, type 2 diabetes and certain cancers to recommend moving to more plant-based diets. In this case, both human health and the environment will benefit.

Glyn Davies and Duncan Williamson demonstrate that, in the UK and the European Union (EU), WWF is deeply engaged in promoting more sustainable eating patterns. What does that mean in practice? Of course there is synchronicity with the Harvard authors: eat less meat, eat more plants.

Neurologist Aysha Akhtar gives us a worrying scenario regarding the likelihood of a global pandemic of avian influenza or a similar virus. As she points out, the average chicken shed, with over 20,000 birds closely packed together, is an ideal hotbed in which viruses can – and do – mutate. It may be just a question of time…

If the obvious question seems to be "Why have animal farming at all?", then Andrew Knight and Jasmijn de Boo claim to have the answer and make a strong case for veganism.

As we all know, money talks – at least it certainly influences. Rosie Wardle urges investors to invest not in antibiotic-dependent factory farming but in long-term solutions, such as the alternative "meats". *Forbes* magazine contributor Michael Pellman Rowland sees huge opportunities for companies to invest in the future of food – not the cheap and nasty kind but the sustainable, plant-based foods of the future.

Leah Garces gives encouraging examples of how food business change in the US is benefiting the lives of farm animals. Bruce Friedrich explains how some of the largest, most powerful meat companies are now buying stakes in the alternative meat companies. Change really does seem to be on the way!

We need to be optimistic but realistic. The money, power and influence of agribusiness will not yield to our arguments without a strong defence of the status quo. They will tell us how important it is that the poor can afford to eat, say, fast food burgers or chicken and chips. But we believe that the ground on which they stand is getting increasingly shaky.

The public health evidence for reducing meat consumption and eating more plant-based diets is becoming overwhelming. The impact of livestock production on climate change, biodiversity loss, soil health and water use and water pollution is becoming recognised as a major issue which needs to be addressed urgently.

Compassion in World Farming ended the conference with a call to all our friends and allies to partner with us to achieve change, working with food businesses, investors and farmers, and working at local, national and international levels. We plan to get the United Nations General Assembly on side. That will take some time – although time is running out.

Our message to the conference – and to you, our readers – is this: let's work together to create urgent change to our current food and farming systems. Let's moderate our own diets with that big picture in mind. Let's write those lobbying letters. Let's support organisations striving for change. Let's hasten to bring about a fairer, more compassionate and planet-friendly food system.

## Note

* Other valued conference partners were The University of Winchester, BirdLife International, the Alliance of Religions and Conservation (ARC) and the European Environmental Bureau (EEB). Huge thanks to all our partners.

# Part I

# Setting the planetary scene

# 2 Livestock and the boundaries of our planet

*Katherine Richardson*

At the inception of Compassion in World Farming (Compassion) 50 years ago, very few people were concerned that human activities might influence the Earth at the planetary level. Although Compassion's founding documents reveal a real concern for earth systems such as soil and water, the organisation decided to begin with animal welfare as a primary concern. In the intervening years, a much greater understanding of the Earth as a system, where the state of the system is a function of interactions and feedbacks occurring within system components, has developed. With the Amsterdam Declaration (Steffen et al., 2004) by the Earth System Science Partnership in 2001, it was recognised that the "Earth System behaves as a single, self-regulating system comprised of physical, chemical, biological and human components". This was particularly important as it represents an acknowledgement of the fact that human activities can operate as a "quasi-geological" force in the control of planetary processes.

At the level of individual ecosystems (i.e. lakes, coral reefs, grasslands) it has, of course, long been recognised that both non-human forces and human activities can cause a state change of the ecosystem as a whole (clear to turbid lakes, pristine coral reefs to reefs covered by algae; grasslands to shrub lands, etc.). The Amsterdam Declaration essentially acknowledged that it is not only natural forces but also human activities that have the potential to change the state of the Earth System as a whole. As our understanding of Earth System functioning has evolved, it has become clear that farming – and not least of which livestock production – has an enormous effect on the planet.

This focus on the Earth as a self-regulating system actually began around the same time that Compassion started, when Apollo astronauts took a photo of the Earth from space. For the first time, it became abundantly clear to scientist and non-scientist alike that the Earth is not connected to any other celestial body. What this means, of course, is that the natural resources we take from the Earth and upon which we, ultimately, are dependent are finite. Once they are depleted, they will not be replaced. Currently, we (seven billion people) are sharing these resources with all other living organisms. The United Nations predicts that the global human population will reach nine to ten billion by 2050 and possibly even higher by 2100. The UN Agenda 2030 (Sustainable

Development Goals) adopted in 2015 (www.un.org/sustainabledevelopment/sustainable-development-goals/) provides a vision for how these resources should be shared, but, before we can consider *how* resources are to be shared, we need to have an estimate of how much of a given resource is actually available for sharing.

Through the study of the Earth as a system ("Earth System science"), critical resources can be identified, and estimates of their availability can be developed. One approach for doing this is the *planetary boundaries framework* (Rockström et al., 2009a, 2009b; Steffen et al., 2015). Briefly, this framework is based on the observation that Earth has throughout its history been in numerous different "states", that is, periods with varying environmental conditions, where temperatures have been both warmer and colder than they are today. During the past ~12,000 years, however, the climate conditions have been relatively warm and stable compared to those of the last million years. While humans have been a part of the Earth System for the past ~200,000 years, it is only within these last 12,000 years that everything we associate with modern civilisations (agriculture, written language, etc.) has developed. We know, then, that humanity can thrive in Holocene-like conditions. We do not know for certain that it can thrive under other than Holocene-like states. It would, then, be very unwise for humanity to perturb critical Earth System processes to the point that there is an increase in the risk of the Earth transitioning to a different state.

In developing the planetary boundaries framework, the authors identify nine global processes that are critical for maintaining the Earth System in its current state and are heavily impacted by human activities (Figure 2.1). By studying the variability of these processes throughout the Earth's history, the authors of the framework propose, for seven of the processes, a "boundary" for human perturbation, beyond which they argue that the risk of initiating a change in the Earth System as a whole becomes increased. These "boundaries" should not be confused with absolute thresholds or tipping points but, rather, can be compared to blood pressure measurement. When blood pressure is above 120/80, there is no *guarantee* that a cardiac event will occur, but the risk of an event is increased. Therefore, we attempt to reduce blood pressure. This analogy actually fits quite well when we consider the historical development of human influence on the ozone hole. The planetary boundaries analyses show that, in the 1990s, human impacts on the ozone hole were on the wrong side of the boundary. Today, thanks to actions agreed upon in the Montreal Protocol, human impacts on the ozone hole have been brought within the planetary boundary.

The planetary boundary framework argues, then, that the region within the ring designating the planetary boundaries in Figure 2.1 represents a "safe operating space" for humanity. In other words, if we are able to constrain our impact on these critical global processes to within this ring, the framework predicts that there will not be a substantially increased risk that anthropogenic forcings will drive the Earth System to a different state. Currently, the

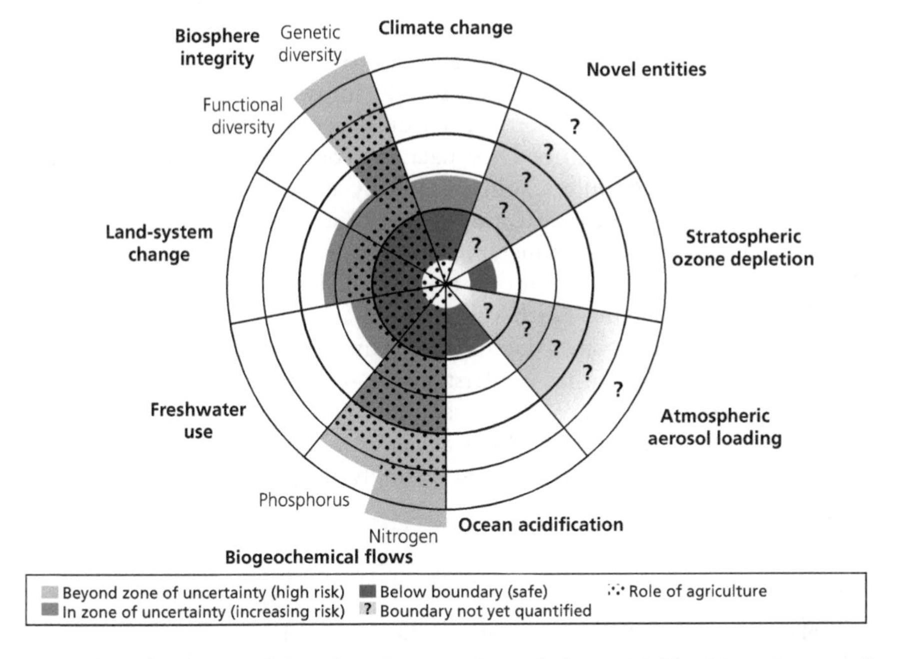


*Figure 2.1* The status of the nine planetary boundaries overlaid with estimates of agriculture's role in that status.

Sources: Steffen et al. (2015) and Campbell et al. (2017).

Note: that the position of the water planetary boundary denoted by Steffen et al. (2015) has been modified by Campbell et al. (2017).

authors (Steffen et al., 2015) of the framework argue that we have crossed four of the nine boundaries: climate, biosphere integrity (biodiversity), land use (felling of forest) and the release of reactive nitrogen and phosphorous to the environment.

Campbell et al. (2017) have estimated the contribution of agriculture, including livestock farming, to the human perturbation of the global processes included in the planetary boundaries framework (dotted regions superimposed on the boundaries depicted in Figure 2.1). Agriculture makes a significant contribution to the climate boundary, but the impacts of agriculture on the climate system are, in themselves, not enough to bring human perturbation beyond the planetary boundary. With respect to the impacts of livestock farming on climate it has been estimated that ~8%–18% of global greenhouse gas emissions are coming from livestock (O'Mara, 2011; FAO, 2006, 2013; Herrero et al., 2013).

For the other three processes where Steffen et al. (2015) find the planetary boundary to be exceeded, the influence of agriculture alone is sufficient to bring the level of humanity's perturbation beyond the proposed boundary, that is, outside of the "safe operating space". Thus, these studies indicate that substantial changes in the global agriculture/food system are necessary in order to minimise the risk of humanity inadvertently transitioning the Earth System to a different state.

The analysis conducted by Campbell et al. (2017) treats agriculture generally, and it is not possible to directly assess the impact of livestock farming on the planetary boundaries. Nevertheless, from sheer numbers alone (according to The Food and Agriculture Organization Corporate Statistical Database (FAOSTAT), 2016, 20 billion animals globally at any one time), we can expect the impact of livestock farming to be substantial. With respect to the exceedance of the land use planetary boundary, it can be noted that Herrero et al. (2013) estimate that about one-third of the ice-free area of the Earth is used for livestock systems. Furthermore, Nepstad et al. (2011) attribute ~72% of deforestation to the conversion of land to livestock use.

Thus, developments in the understanding of impacts of agricultural activities on Earth System processes strengthen Compassion's call for a major transformation of the industrialised agricultural system. Indeed, sustainable development of human societies is simply not possible without such a transformation!

## References

Campbell, B. et al. (2017) 'Agriculture production as a major driver of the Earth system exceeding planetary boundaries', *Ecology and Society*, vol 22, no 4, article 8.

Food and Agriculture Organization of the United Nations (2006) Livestock's Long Shadow – Environmental Issues and Options. FAO, Rome.

Food and Agriculture Organization of the United Nations (2013) Tackling Climate Change through Livestock – A Global Assessment of Emissions and Mitigation Opportunities. FAO, Rome.

FAOSTAT (2016) The Food and Agriculture Organization Corporate Statistical Database, Food and Agriculture Organization of the United Nations (FAO), Rome.

Herrero, M. et al. (2013) 'Biomass use, production, feed efficiencies, and greenhouse gas emissions from global livestock systems', *Proceedings of the National Academy of Sciences of the United States of America*, vol 110, no 52.

Nepstad, D.C. et al. (2011) 'Systemic conservation, REDD, and the future of the Amazon Basin', *Conservation Biology*, vol 25, no 6, pp. 1113–1116.

O'Mara, F.P. (2011) 'The significance of livestock as a contributor to global greenhouse gas emissions today and in the near future', *Animal Feed Science and Technology*, vol 166–167, pp. 7–15.

Rockström, J. et al. (2009a) 'A safe operation space for humanity', *Nature*, vol 461, pp. 472–475.

Rockström, J. et al. (2009b) 'Planetary boundaries: Exploring the safe operating space for humanity', *Ecology and Society*, vol 14, no 2, article 32.

Steffen, W. et al. (2004) *Global Change and the Earth System: A Planet under Pressure*. Springer, Heidelberg.

Steffen, W. et al. (2015) 'Planetary boundaries: Guiding human development on a changing planet', *Science*, vol 347, no 6223.

Part II

# The impacts of livestock production on the natural world

# 3 The great disappearing act

*Philip Lymbery*

## Why factory farming is a major driver of wildlife declines

What on Earth have elephants, jaguars and penguins got to do with factory farming? After all, isn't it all about cruelty to pigs, chickens and cows?

Well, the truth is that factory farming to produce "cheap" meat is not only the biggest cause of animal cruelty on the planet but also a major driver of wildlife declines. It causes one of the greatest disappearing acts in the natural world, putting at risk the bedrock of future society: food.

In recent years, through my work for Compassion in World Farming, I have travelled extensively to investigate what's really going on in the food system.

My forays have taken me to many colourful places, perhaps none more so than the Indonesian teardrop island of Sumatra. I went there to investigate a little-known facet of the palm industry, namely the use of palm products for animal feed. Deforestation to make way for intensive palm plantations is destroying the last habitat of the orangutan, Sumatran tiger, rhino and elephant. I discovered that large quantities of palm kernel, the edible nut from the trees, as well as oil, are being shipped out to feed intensively farmed cattle and other animals, particularly in the European Union (EU). This boosts the profitability of the palm industry, encouraging further deforestation, causing more species-rich jungle to be cleared for sterile plantations. In the case of the elephants, their shrinking habitat is leading to increased conflict between elephants and people, leading to sharp declines in a remarkable creature down to its last 2,500 individuals.

Sumatra's palm industry is highly protective of what it does; despite warnings of hostage-taking, shootings and arrests, I decided to press on with the investigation. Everywhere I went I was followed by a dull, hazy smog. So bad that it even disrupted air traffic and caused the government to declare a state of emergency (McCafferty and Sater, 2015).

This smog was coming from rampant deforestation; burning forests to make way for more industrial palm plantations, adding to climate change, and leading to the demise of the critically endangered Sumatran elephant (Figures 3.1 and 3.2).

*Figure 3.1* Factory farming is driving iconic wildlife to extinction including the Sumatran elephant. Photo credit: Philip Lymbery.

*Figure 3.2* Deforestation for palm plantations is destroying the habitat of the critically endangered Sumatran elephant.

Just months later, media coverage brought the welcome news that the world's governments had met in Paris to agree a groundbreaking United Nations (UN) Convention to stop runaway climate change. The world breathed a sigh of relief.

Yet my recent trip showed how tightly interwoven environmental issues can be and begged the question: if we could solve climate change without fixing the industrial food system, would life on Earth be safe? Would the future be secure for our children?

Well, from what I'd learned, the answer is an emphatic "no" because there's another major challenge facing humanity; just as serious and with consequences that are far more permanent; and it's on our plate.

## Spreading deserts of green

I've been a wildlife enthusiast since I was a small boy. When I was 8 years old I got chicken pox and, to cheer me up, my mother gave me the Observer book of birds. My mother was delighted; I entertained myself for weeks off school by watching the birds in our garden.

That moment fired a life-long passion. So much so, that in my teens, I'd be off to the forests, the farmland and reservoirs to watch birds. I was absolutely fascinated by birds and their power of flight, by migration, by how they embodied freedom.

My mid-teens brought another life-changing moment. Someone came to talk at my school from an organisation I'd never heard of before: Compassion in World Farming.

He put on a video called *Don't look now, here comes your dinner*, which showed tens of thousands of chickens crammed into windowless sheds; pigs in crates unable to turn around for weeks and months at a time; cattle crowded in feedlots where they are fed corn, not a blade of grass in sight. Yet the thing that really got me was seeing hens in battery cages; birds kept their entire life in cages so small they couldn't even flap their wings. As a wildlife enthusiast fired by a sense of freedom and the power of flight, it blew my mind.

What I didn't realise back then was that it wasn't only farm animals that were disappearing from the land; the wild creatures in the British countryside which so fascinated me were disappearing before my very eyes too. In my lifetime, Britain has lost 44 million birds; that's one breeding pair gone every minute (RSPB, 2012).

In Britain and elsewhere, once-common farmland species have gone into steep decline. Birds such as turtle doves, grey partridges, corn buntings and tree sparrows have declined by 90% or more over the last 40 years. The skylark, lapwing and even the common starling have dwindled by at least 60%. In recent decades, two million pairs of skylarks and a million pairs of lapwings have simply disappeared.

These declines are not confined to Britain: European bird census results for 1980–2010 show that "farmland birds have fared particularly badly", with 300 million fewer birds today than in 1980. Grey partridge and crested lark have been hit particularly hard, with declines of more than 90%. Ortolan buntings, turtle doves and meadow pipits have seen their numbers slashed by more than two-thirds (EBCC, 2014).

In the US, where farmland birds are called "grassland" or "shrubland" birds, many species are also in deep trouble. Those suffering include eastern meadowlark, lark bunting, mountain plover, short-eared owl and burrowing owl (Sauer et al., 2014).

Again, industrial farming is largely to blame.

Birdlife International says the declines in Europe are widely accepted as being driven by agricultural intensification and the resulting deterioration of farmland habitats (BirdLife, 2013). US farmland bird losses are thought to be in response to the loss of small farms, declines of shrub habitat and expanding "industrial agriculture" (Murphy, 2003).

Industrial agriculture has since swept the landscape in the UK, Europe, the US and beyond, leading to widespread declines in wildlife and the diversity of nature. It has also been exported across the world, not least, to Asia and South America.

In the last 40 years, since the widespread adoption of factory farming, the total number of wild mammals, birds, reptiles, amphibians and fish worldwide has halved (Bringezu et al., 2013; McLellan et al., 2014; Owen, 2005).

Much of this decline comes down to the two sides of factory farming: side one is where the farm animals are kept; taken off the land, out of fields and pastures into cages and crates. This looks like a space-saving idea, but actually isn't. Confining farm animals means their feed has to be grown elsewhere – the second side of factory farming – on scarce arable land, usually using chemical pesticides and fertilisers.

And in so doing, it sparks off a great disappearing act.

It's not just farm animals that disappear from the land but also the trees, the bushes and the hedges, along with wild flowers. And when they disappear, so too do the insects and the seeds needed by the birds, the bats, the bees and so much else. The worms beneath our feet disappear, along with soil fertility; leaving little else but the crop.

Then, in perhaps the greatest disappearing act of all, we take this crop and feed it to factory farmed animals, who waste most of the food value of that crop, be it in terms of calories or protein, in conversion to factory farmed meat, milk and eggs. Studies show that some two-thirds of the potential food value or calories of grain are lost (Lundqvist et al., 2008; Nellemann et al., 2009). Beef has the worst conversion rate. According to a study by the University of Minnesota, for every 100 calories fed in the form of grain, as little as 3% is returned in the resulting meat (Cassidy et al., 2013). The picture in relation to protein is little better. The same study found that for every 100 grams of protein fed to farm animals in the form of grain, we get back only about 43 grams of protein from milk, 40 from chicken, 35 from eggs, 10 from pork and 5 from beef.

The UN Food and Agriculture Organisation (FAO) puts it like this: "When livestock are raised in intensive systems, they convert carbohydrates and protein that might otherwise be eaten directly by humans and use them to produce a smaller quantity of energy and protein. In these situations, livestock can be said to reduce the food balance" (FAO, 2011). See Peter Stevenson's chapter in this book for more detail on this vital issue.

The phrase "reduce the food balance" is political speak for "wasted". It's madness on a plate.

## Rainforest, savannah and soy

It was perhaps in Brazil where I came to appreciate the sheer scale of what's going on. I was there on the trail of the jaguar, a species down to its last 15,000 individuals, half of them in Brazil.

I visited Emas National Park, a rare pristine patch of Cerrado, the savannah grassland that once covered much of this interior landscape.

I remember it being like nature's last stand: on one side of the road, a rich web of life; on the other, a vast prairie of uniform farmland. There were no hedges, no field boundaries, just one newly planted patch of crop after another: soya.

I asked a leading biologist known as the "dean" of jaguar conservation, Leandro Silveira, what all this means for the big cats: "When you deforest the land and remove their prey species, even if you're doing farming of soya bean and corn... you're directly affecting the cat... You are silently wiping out the species".

During the same trip, I took a flight from Sao Felix, a place the locals call "the end of the Earth". I flew in a small plane at 10,000 feet, where you see over such a distance, it feels like you can see the curvature of the Earth. I flew over endless rainforest. I saw the mighty Araguaia River wriggling away like a giant serpent.

As I flew further south, things started to change; the rainforest started to show bare patches; just small ones at first. Then bigger. Then great chunks. Until all of a sudden, the rainforest was but an island in this vast sea of crops, then it was gone...

I watched, transfixed over what was now a never-ending crop prairie of soya. There were no trees or hedges in sight. No relief from the relentless crop-filled landscape. What I did see were clusters of half a dozen or so massive combine harvesters sweeping across the landscape like aerobatic teams. This was on a scale like nothing I'd seen before.

I had just watched the lungs of the Earth disappearing. So, what is all this soya for? Is it to feed the veracious appetite of rampant vegetarians? No. It's to feed factory farmed animals, many of them in Europe.

Soya is actually a wonder crop, one of only a handful of plants that provide a complete protein, containing all the essential amino acids needed for human nutrition (Soyatech, no date). Yet, despite its undoubted nutritional value and the devastation wreaked on vast swathes of the landscape to produce it, only a fraction actually goes to feed people.

The vast majority is used for animal feed. Most of the soya beans (85%) are crushed to give oil and soya meal. Oil makes up less than a fifth of the pulped beans and largely goes for vegetable oil, with small amounts finding their way into soaps or biodiesel. Nearly four-fifths of the pulped bean become soya meal destined for the feed troughs of intensively farmed animals like pigs, chickens and cattle.

When defenders of factory farming suggest that cramming animals into airless barns "saves space", they fail to take into account that the business model is wholly dependent on large amounts of space elsewhere. That space and associated environmental damage are effectively outsourced, often to another continent. The EU imports about 35 million tonnes of soya every year, almost half coming from Brazil. Some 13 million hectares of South American land – an area roughly equivalent to a country the size of Greece – is dedicated to growing soya for the EU, much of it to feed industrially reared farm animals (WWF, 2014).

In Brazil, the soya industry is expanding by hundreds of thousands of hectares every year, largely by ploughing up already deforested, cattle pastures. In the state of Mato Grosso between 2011 and 2016, soya production increased by a third (IMEA, 2016), with industry sources suggesting it could double again (Stewart, 2015).

Yet ploughing up existing cattle pastures for soya doesn't avoid deforestation. Researchers found that, in Mato Grosso, conversion of pasture to soy production has displaced cattle farmers further north into the forests, where they fell trees to create new grazing land. So, while clearing forests for new cattle pasture may be the direct reason for deforestation, the expansion of soybean production across pasturelands is arguably the major underlying cause (Barona et al., 2010).

If anything, the spread of soya by ploughing up cattle pastures is having an amplifying effect on deforestation elsewhere. It makes it lucrative for cattle farmers to sell existing pasture to soya producers and move somewhere cheaper, usually into the forest. In the early 2000s, the global soya boom caused land prices in the Amazon to rocket. In some parts of Mato Grosso, values shot up tenfold. The land rush allowed cattle ranchers to sell their fields for enormous profits and expand their herds by buying land further north. Newly (and illegally) deforested land was relatively cheap (Barona et al., 2010).

American scientists describe this process as a "land-use cascade". They calculate that if soya expansion were reduced, its effect on the amount of rainforest saved would be disproportionately large. According to the study, a 10% reduction in soya expansion into old pasture areas between 2003 and 2008 would have reduced Amazon deforestation by as much as 40% (Arima et al., 2011). This reflects rising land values and the fact that cattle farmers displaced by selling their pasture for soya can buy – and therefore clear – so much more land in the Amazon.

As soya spreads across the Brazilian landscape, seeing a jaguar becomes ever harder; so I headed for the Pantanal, the largest continental wetland on the planet and one of the best places in the world to see them in the wild. I tingled with anticipation as I set off along the hundred-mile Transpantaneira road, a dirt track barely worthy of the name "road". I crossed more than 120

*Figure 3.3* Brazil is home to half of the world's remaining 15,000 jaguars.

bridges – many of them makeshift, alarmingly so – until the road petered out beside the Cuiaba River at the tiny settlement of Porto Jofre (Figure 3.3).

Whilst at Porto Jofre, I couldn't help but wonder how many people realise that the reason for the forest and savannah disappearing – and wildlife with it – is to feed factory farmed animals, often on other continents.

The bitter truth is that cheap meat in Britain, Europe and elsewhere – whether it be pork, beef or chicken – is likely to have been reared on soya from the deforested plains of South America; something that really shouldn't be lost on anyone remotely conservation-minded.

As I settled in for my one night on the Pantanal, the hotelier brought me a booklet on the region's jaguars. The glossy centrefold featured a picture of some of Porto Jofre's tourists photographing the big cats. They were keen wildlife enthusiasts, perhaps even conservationists, just like me. They descend here from all over the world: the US, Italy, Germany and the UK. As I looked intently at that picture of eager wildlife enthusiasts like me, I couldn't help but wonder how many arrive stoked up on cheap meat from soya-fed animals.

## The power of pasture

My investigative journey took me to the Midwest of America, to the Great Plains, where once upon a time you could find the most supreme example of the power of pasture: bison.

Caught in the crossfire between the European settlers and indigenous Native Americans, bison were all but wiped out. In the eyes of the authorities, as one commander put it, "every buffalo dead is an Indian gone".

Wild bison were saved from extinction by the establishment of Yellowstone National Park in 1872, America's first and still largest national park. It marked a new era of concern for the natural world.

I arrived in Yellowstone to see wild bison and I only had one day to do it. Yellowstone is massive, covering nearly 3,500 square miles. And although bison are pretty big, they're hard to find in an area that size. So, I set off before dawn and drove for hours. Finally, just south of a village called Canyon, the moment I'd been waiting for: there, grazing contentedly beside the car, were two bull bison so close I could hear them munching. Steam streamed from their backs in the early morning sun, and puffs of breath dissolved in the cool air as they focussed on the serious business of breakfast. They had strongly triangular outlines: tapered at the back and big at the front, accentuated by that characteristic hump at the shoulders. Their heads were unfeasibly broad, their coats matted and curled. It was a formidable sight.

It put me in mind of how things were up to the late 1800s when 30–50 million bison roamed the Great Plains of America. They weighed in total about the same as the entire human population of North America today. Always on the move, these vast herds were sustained by nothing more than rain, sunshine and grass. They were perhaps the most potent example of grazing animals living harmoniously with their environment; of the power of pasture.

Few wild bison roam freely on these plains today. The ones I saw in Yellowstone were grazing in pasture clearings surrounded by mountains and trees. An impressive sight, yes, but a pale shadow of what used to be. The park itself is surrounded by farmland; these natural wanderers confined to the sanctuary like refugees. If they leave, they're shot. If their numbers swell beyond a few thousand, they are rounded up and culled.

Yellowstone's bison have become an example of what I call "big-country claustrophobia", where wide-open spaces are no longer big enough to sustain charismatic megafauna. I saw this in South Africa's national parks, where vast wilderness supports a surprisingly small number of large animals like elephants or lions. I saw it here in Yellowstone, where relatively small numbers of bison – a tiny fraction of the original herd – remain under constant threat of being culled.

My time in the park was all-too-brief; I was soon heading southwards to the Great Plains as they are now. I clambered into a huge tractor with as much technology as a space shuttle. From this elevated position, as far as the eye could see was what the farmer beside me described as a "flowing green ocean of corn".

I asked him what this vast genetically modified prairie of corn was all about: "Feeding the world", he said; yet most of the corn was destined to feed cattle and cars. Indeed, more than a third of the entire cereal crop globally goes to feed industrially reared animals in a process so inefficient it gives back far less calories and protein than it consumes. In this way, cereal crops that are enough to feed three billion extra people are wasted.

I was struck by the irony that, like the bison, many cattle no longer have the freedom to roam. In the Midwest state of Nebraska, the signs suggested this was "cowboy" country, but there was no need for Stetsons on horseback, no roundups required. These were feedlots rearing "battery" beef.

A thousand cattle stood motionless in muddy paddocks, not a blade of grass in sight. You'd have thought it would be noisy, but instead there was an eerie hush. It was like a hospital ward, the quiet broken only by the odd cough, sneeze or wail. The stench of excrement was overpowering. Full grown cattle and tiny calves were standing in the fierce Nebraskan summer sun with no shade, desperately trying to lie in each other's shadow.

Those cattle will be fed corn most likely grown in chemical-soaked monocultures on prairies that, before the plough, once represented sustainable grasslands.

Enormous amounts of fertiliser made in chemical plants the size of an industrial estate are used to keep tired soils producing, yet much of it washes into waterways and down the Mississippi where it ends up in the Gulf of Mexico. There, it causes one of the world's largest marine dead zones; an area of polluted sea as big as a country the size of Wales, where nothing lives. The zone emerges every year, without fail, from February to October, stretching all the way from the shores of Louisiana to the upper Texan coast (EPA, 2018).

It is a lifeless bottom layer of ocean where the oxygen is sucked out, driving anything alive towards the surface. As the dead zone spreads, some bottom-dwelling fish are forced to the surface, where they are vulnerable to predators; some may flee the area; and the rest just die.

The US government's scientific agency, the National Oceanic and Atmospheric Administration (NOAA, no date), estimates the dead zone to cost US seafood and tourism industries' $82 million a year, a significant blow to the Gulf Coast economy (McKinney, 2014). Despite the economic cost, the problem continues to escalate. In 2015, the dead zone covered 6,474 square miles, three times bigger than the reduction target set by an official task force in 2001 (NOAA, 2015). And the main culprit? Fertiliser.

According to the NOAA, the source of the problem are those "flowing green oceans" of corn I saw in Nebraska and across much of the Midwest. In a report entitled *The Causes of Hypoxia in the Northern Gulf of Mexico*, the NOAA describes this part of the US as "an area of intensive corn and soybean production, where large amounts of nitrogen from fertiliser and manure are applied to soils every year". The agency report explains how excess nitrate is washed into rivers and streams, and ends up in the Gulf, the scientific evidence for which is described as "overwhelming" (NOAA, no date).

Dead zones are now emerging around the world. Since the 1960s, the number of dead zones worldwide has almost doubled every decade (VIMS, 2018a). There are now more than 400 coastal dead zones in the world, affecting a total area of some 95,000 square miles – about the same size as New Zealand (VIMS, 2018b). Most are found in temperate waters, off the eastern coast of the US and in the seas of Europe. Some are brewing in the waters off China, Japan, Brazil, Australia and New Zealand. The world's largest dead zone is in the Baltic Sea, where nutrient-enriched runoff from farms has combined with nitrogen deposition from the burning of fossil fuels and human waste discharged directly into the sea (Dybas, 2005).

For now, at least, dead zones are reversible; if farmers and policymakers can be persuaded to stem the flow of nutrient pollutants into lakes and coastal areas, then things can be turned around. However, as the Gulf of Mexico illustrates, trying to change things in the face of powerful agricultural interests is a struggle. All too often, governments seem to prefer an approach of out of sight, out of mind.

## Oceans plundered

Threats from industrial agriculture to the complex web of oceanic life come not only from pollutants put into the sea but also from what is taken out.

Vast quantities of fish are caught, not to feed people, but to be ground down into fish meal to feed factory farmed fish, chickens and pigs, leaving wildlife like penguins and puffins starving.

I came to realise the true extent of how factory farming plunders the ocean during a trip to South Africa's Boulders Beach along the Cape Peninsula. Here, a colony of African penguins have recently set up home near to residential houses (Figure 3.4).

A hundred years ago, African penguins were one of the most abundant seabirds in the region, numbering three to four million birds. Now, the penguin population is estimated to be 80,000 individuals (BirdLife International, 2016) and falling fast.

*Figure 3.4* The African (or jackass) penguin could become extinct within the next 15 years.

A display board on Boulders Beach is clear about the threats facing the birds, which it states include the "reduction of penguin food supply by commercial fishing". South Africa's marine ecosystem relies on three small pelagic fish for food: anchovies, sardines and red-eyes. These species drive the rest of the ocean community, from the hake and yellowtail fish that eat the small pelagic fish to the sharks and tuna that, in turn, eat those fish to penguins, seals, dolphins and whales. They all depend on the little fish, known in fishery circles as forage, bait, prey, or "trash" fish. Now they are running out.

The UN warns that South Africa's anchovy and sardine (pilchard) fisheries are being overexploited (Hecht and Jones, 2009).

Dr Lorien Pichegru, a leading marine biologist based at the Institute of African Ornithology, fears that within 15 years, the African penguin could suffer the same fate as the dodo: "Overfishing is a huge concern", she said. "Both fishermen and penguins are struggling because the fish population is also declining. It is very worrying" (Memela, 2013).

Yet this is a problem far from confined to southern Africa; worldwide, over 17 million tonnes of small pelagic fish, like anchovies, sardines, herring and sandeels, are removed from the ocean every year – an estimated 90 billion individual animals. This accounts for nearly one-fifth of all the marine fish catch globally, much of it for fish meal.

Worldwide, the fish used in this way would be enough to provide a billion more people with a dietary supply of fish. Or, put another way, leaving them in the ocean would take huge pressure off hard-pressed fish stocks.

Seabirds are far from the only victims of plundering the world's oceans to feed factory farms: human food supplies also suffer. In the last half-century, about 90% of the world's big fish – those we put on our plates – have been taken for food or discarded, leaving oceans close to collapse (Earle, 2009).

Scientists predict that most of the world's fisheries will be depleted by 2048 (Worm et al., 2006).

Those small fish targeted for rendering into fish meal – herring, anchovies, sandeels and the like – are a vital part of the food chain: the essential ecological link between microscopic phytoplankton and larger fish like cod and tuna. Put simply, the fish we eat rely on the small pelagic fish they eat for their survival.

## The stuff of life

Nature's great disappearing act is playing out across the world's bays and oceans, forests and grasslands, to the detriment of both farm animals and wildlife.

It also affects resources imperative to the future well-being of humanity.

Antibiotics are disappearing: half the world's antibiotics are fed to farm animals, much of it to ward off diseases of intensification. In the US, farm-animal use can be as high as 80% of total antibiotic use. In Britain, nearly 90% of farm antibiotics go to poultry and pigs, the species most intensively farmed

(VMD, 2013). Antibiotic use in livestock is acknowledged by the World Health Organisation (WHO) as a factor in the rise of so-called superbugs: bacteria resistant to antibiotic treatment. The WHO fears the world may be on the cusp of a post-antibiotic era where once-treatable diseases could once again kill.

Bees are also disappearing, despite our reliance on them to pollinate a third of the world's crops. Numbers have almost halved in the past 25 years in England (BBC News, 2013). Researchers at the University of Reading believe that Britain has less than a quarter of the bees needed for the proper pollination of crops, while Europe has only two-thirds. "If these wild bee populations collapse there would be nothing to compensate for them", the University of Reading's Dr Tom Breeze told *Farmers Weekly* magazine (Jones, 2014).

Perhaps most importantly, soils are disappearing too. According to a report by government advisory body, the Committee on Climate Change (CCC), declining soil health and erosion mean that large areas of farmland in the UK are in danger of becoming unproductive within a generation (Bawden, 2015; Global Agriculture, 2015). As a result, coming decades could see Britain's ability to produce food decrease in the face of increasing demand. Farmers have been able to improve yields through technological advances, but mounting evidence suggests that it is only a matter of time before productivity declines.

Lord Krebs, chairman of the CCC's adaptation subcommittee, said, "Soil is a very important resource which we have been very carefree with. At the moment we are treating our agricultural soils as though they are a mined resource – that we can deplete – rather than a stewarded resource that we have to maintain for the long-term future" (Bawden, 2015).

Intensive agriculture is directly responsible for the problem, according to the CCC report, which says that "deep ploughing, short-rotation periods and exposed ground" is leading to soil erosion from wind and heavy rain (Committee on Climate Change, 2015).

The figures are stark. Since 1850, Britain has lost 84% of its fertile topsoil, with erosion continuing at a rate of 1–3 cm a year. Given that soil can take hundreds of years to form (Syngenta, no date), these losses are not sustainable. Krebs has warned that the most fertile topsoils in the east of England – where 25% of potatoes and 30% of vegetables are grown – could be lost within a generation (Committee on Climate Change, 2015).

Soil is the foundation of life on Earth. Without it we can't grow food. Yet industrial agriculture treats it with disdain.

A recent UN report by 200 soil scientists in 60 countries concluded that the condition of most of the world's soils is fair, poor or very poor, and is getting worse. According to the FAO, at current rates of depletion the world's topsoils could be gone within 60 years (Arsenault and Russell, 2014). At that rate, a child born today won't even reach retirement age before he or she perhaps witnesses the death of our soils, the end of the food system as we know it.

Which brings me back to my earlier question: if we could solve climate change without fixing the industrial food system, would life on Earth be safe? Would the future be secure for our children?

Well, clearly not.

## Welcome to the Anthropocene

The loss of our natural world is a crisis at least as big, and far more permanent, than climate change. Scientists suggest that we have entered a new geological era, the "Anthropocene", a new age in which *Homo sapiens* are inflicting wholesale and irreversible changes to the planet. If we simply carry on as before, scientists warn of a mass extinction, perhaps the biggest since an asteroid wiped out the dinosaurs. Species are already disappearing at a rate 1,000 times higher than previously expected.

About two-thirds of the overall loss of wildlife are driven by food production (Secretariat of the Convention on Biological Diversity, 2014). Industrial agriculture – factory farming – is the most damaging.

As forests are felled, rivers polluted and soils eroded, ecosystems collapse; putting at risk the air we breathe, the water we drink and the food we eat.

When all else fails, defenders of industrial agriculture fall back on that tired argument about "feeding the world": that we need to increase, even double, food production to feed a growing human population. When agriculture already covers half the usable land surface of the planet, there is little room for growth.

Yet the real challenge is to recalibrate the world's food system to be around feeding people rather than the industrial production of commodities to feed landfill, cars and factory farms.

Today, there are seven and a half billion people, soon to be ten. According to UN data, the world already produces enough food to feed up to 16 billion people; that's more than enough for everyone today and in the foreseeable future.

The problem is that much of it is wasted; and the biggest form of food waste on the planet is in feeding human-edible crops to industrially reared animals. In this way, we waste enough food to feed an extra four billion people on the planet. That's not to say that four billion extra people is a good idea: it is not. What it shows is that *without* factory farming, we could feed everyone with better quality food using less farmland, not more. Leaving room for nature. Allowing the planet to breathe. Preserving fresh air, clean water and soils for future generations.

## The great reappearing act

Wouldn't it be great if we had a farming system that, instead of being wasteful, inefficient and the single biggest cause of animal cruelty on the planet, actually started to put things back? That started to regenerate the soil naturally?

That started to bring pollinating insects swarming back? That didn't rely on copious antibiotics? That made wonderful high-welfare environments for farm animals and for the wild things too?

What if farming were *regenerative*?

Well, the good news is that it can be and, for some farmers, still is.

Let me introduce you to someone who is doing just that: Tim May, who inherited an industrial arable farm in Hampshire, England, ten times the size of the UK average. He used to drench his crops with a battery of pesticides and fertilisers: nine different chemical applications every harvest.

He noticed his soils were becoming tired and yields starting to dip.

He decided to dispense with the chemicals and monocultures and restored farm animals to their ecological niche as grazing animals, moving them around the farm interspersed with crops.

Cows now follow sheep, followed by pigs, then chickens and then crops. It's a harmonious system. May now leaves most of the chemicals in the barn. And saves himself £700 per hectare in the bargain. His cattle and sheep are purely pasture-fed, and his chickens are free range. His soils are recovering, and wildlife, like barn owls, are flooding back.

May is just one of a growing movement of farmers worldwide moving away from factory farming in favour of a more compassionate, regenerative way of producing food. On land-based, regenerative or agroecological farms where soil fertility is built naturally.

The farm animals can run and jump, flap their wings or wallow in the mud; and in the case of pigs and poultry, they can be restored to their natural role as foragers and recyclers of food waste.

Without the chemicals, wild flowers start coming back, along with seeds and insects, and with them, the birds, bats, bees.

And what's really special about all of this is that it takes the pressure off hard-pressed forests. So they can do under-rated things, like taking carbon out of the atmosphere, and giving us oxygen to breathe.

It's a vision of the countryside as it really should be.

It's a great *reappearing* act; one that could help save life on Earth from extinction; and one which we can all play a part.

As consumers, we can take action on our plate three times a day through our food choices; by choosing to eat more plants, less and better meat and dairy from animals that are pasture-fed, free range or organic.

As consumers, we have tremendous power.

Yet to really safeguard the future, we're going to need something more; some kind of global agreement to replace factory farming with a regenerative food system. Something akin to the UN Convention on Climate Change agreed in Paris.

To achieve this, we'll need to mobilise governments, corporates and civil society to bring about a new zeitgeist in favour of a fusion between food, farming and nature. We'll need to act fast as time is running out.

To make this happen, we need to move away from an approach that treats the countryside like an industrial complex to one that works with nature, not against her. And the stakes couldn't be higher. If we fail to act, then we risk undermining the food of future generations. Increasingly, big picture signs are suggesting that we are the last generation that can genuinely leave a planet worth having as a legacy for our children. Moving beyond farming's failed industrial era will bring about a glorious *reappearing* act; a better day for farm animals, for wildlife and for people today and in the future.

## Acknowledgements

The issues in this chapter are explored in my latest book *Dead Zone: Where the Wild Things Were*, Bloomsbury Publishing PLC.

## References

Arima, E.Y., Richards, P., Walker, R. and Caldas, M.M., 2011. Statistical confirmation of indirect land use change in the Brazilian Amazon. *Environmental Research Letters*. 6(2). http://iopscience.iop.org/article/10.1088/1748-9326/6/2/024010/meta. Accessed 2 May 2018.

Arsenault, C. and Russell, R. (Ed.), 2014. 'Only 60 years of farming left if soil degradation continues', *Scientific American*. www.scientificamerican.com/article/only-60-years-of-farming-left-if-soil-degradation-continues/. Accessed 2 May 2018.

Barona, E., Ramankutty, N., Hyman, G. and Coomes, O.T., 2010. The role of pasture and soybean in deforestation of the Brazilian Amazon. *Environmental Research Letters*. 5(2). http://iopscience.iop.org/1748-9326/5/2/024002/media. Accessed 2 May 2018.

BBC News, 2013. What is killing Britain's honey bees. A horizon special, BBC Two. www.bbc.co.uk/news/science-environment-23546889. Accessed 2 May 2018.

BirdLife International, 2013. Europe-wide monitoring schemes highlight declines in widespread farmland birds. www.birdlife.org/datazone/sowb/casestudy/62. Accessed 2 May 2018.

BirdLife International, 2016. The IUCN Red list of threatened species 2016. Spheniscus demersus. www.iucnredlist.org/details/22697810/0. Accessed 2 May 2018.

Bawden, T., 2015. Soil erosion a major threat to Britain's food supply, says Government advisory group. *Independent*. www.independent.co.uk/news/uk/home-news/soil-erosion-a-major-threat-to-britains-food-supply-says-government-advisory-group-10353870.html. Accessed 2 May 2018.

Bringezu, S., Schütz, H., Pengue, W., O'Brien, M., Garcia, F., Sims, R., Howarth, R.W., Kauppi, L., Swilling, M. and Herrick, J., 2013. Assessing global land use: Balancing consumption with sustainable supply. UNEP International Resource Panel. www.unenvironment.org/resources/report/assessing-global-land-use-balancing-consumption-sustainable-supply-0. Accessed 2 May 2018.

Cassidy, E.S., West, P.C., Gerber, J.S. and Foley, J.A., 2013. Redefining agricultural yields: From tonnes to people nourished per hectare. University of Minnesota, *Environmental Research Letters*. 8. http://iopscience.iop.org/article/10.1088/1748-9326/8/3/034015/meta. Accessed 2 May 2018.

Committee on Climate Change, 2015. Progress in preparing for climate change, 2015 Report to Parliament. www.theccc.org.uk/wp-content/uploads/2015/06/6.736_CCC_ASC_Adaptation-Progress-Report_2015_FINAL_WEB_250615_RFS.pdf. Accessed 2 May 2018.

Dybas, C.L., 2005. Dead zones spreading in world oceans. *Bioscience*. 55(7), pp. 552–557. www.academia.edu/5102539/Ocean_Dead_Zones_Spreading_in_World_Oceans. Accessed 2 May 2018.

Earle, S.A., 2009. *The world is blue: How our fate and the oceans are one*. Washington, DC: National Geographic Society, p. 264.

EBCC, 2014. Trends of common birds in Europe, 2014 update. www.ebcc.info/index.php?ID=557. Accessed 2 May 2018.

US Environmental Protection Agency (EPA), 2018. Mississippi River / Gulf of Mexico Hypoxia Task Force. http://water.epa.gov/type/watersheds/named/msbasin/zone.cfm. Accessed 2 May 2018.

FAO, 2011. World Livestock 2011: Livestock in food security. www.fao.org/docrep/014/i2373e/i2373e.pdf. Accessed 2 May 2018.

Global Agriculture, 2015. Soil erosion a major threat to Britain's food supply, warns report. www.globalagriculture.org/whats-new/news/en/30894.html. Accessed 2 May 2018.

Hecht, T. and Jones, C.L.W., 2009. Use of wild fish and other aquatic organisms as feed in aquaculture – a review of practices and implications in Africa and the Near East. FAO Fisheries and Aquaculture Technical Paper No. 518, pp. 129–57. www.fao.org/docrep/012/i1140e/i1140e03.pdf. Accessed 2 May 2018.

Instituto Mato Grossense de Economia Agropecuaira (IMEA), 2016. Soja. www.imea.com.br/upload/publicacoes/arquivos/R404_392_BS_REV_AO.pdf.

Jones, D., 2014. Europe lacks bees to pollinate its crops. *Farmers Weekly*. www.fwi.co.uk/arable/europe-lacks-bees-to-pollinate-its-crops.htm. Accessed 2 May 2018.

Lundqvist, J., de Fraiture, C. and Molden, D., 2008. Saving water: From field to fork – curbing losses and wastage in the food chain. SIWI Policy Brief. SIWI. www.siwi.org/wp-content/uploads/2015/09/PB_From_Filed_to_fork_2008.pdf. Accessed 2 May 2018.

McCafferty, G. and Sater, T., 2015. Indonesian haze: Why it's everyone's problem. *CNN*. http://edition.cnn.com/2015/09/17/asia/indonesian-haze-southeast-asia-pollution/index.html. Accessed 2 May 2018.

McKinney, L., 2014. Louisiana shrimp season threatened by US ethanol policy. NOLA, The Times—Picayune. www.nola.com/opinions/index.ssf/2014/06/louisiana_shrimp_season_threat.html. Accessed 2 May 2018.

McLellan, R., Lyengar, L., Jeffries, B. and Oerlemans, N. (Eds), 2014. Living Planet Report 2014. WWF. http://assets.worldwildlife.org/publications/723/files/original/WWF-LPR2014-low_res.pdf?1413912230&_ga=2.200572776.773180399.1525259355-1953386061.1525259355. Accessed 2 May 2018.

Memela, M., 2013. Penguins facing extinction. *Times Live*, South Africa. www.timeslive.co.za/news/south-africa/2013-02-19-penguins-facing-extinction/. Accessed 2 May 2018.

Murphy, M.T., 2003. Avian population trends within the evolving agricultural landscape of the eastern and central United States. Biology Faculty Publications and Presentations, Portland State University. 70. http://pdxscholar.library.pdx.edu/bio_fac/70. Accessed 2 May 2018.

Nellemann, C., MacDevette, M., Manders, T., Eickhout, B., Svihus, B., Prins, A.G., Kaltenborn, B.P. (Eds), 2009. The environmental food crisis – The environment's role in averting future food crises. A UNEP rapid response assessment. United Nations Environment Programme. www.gwp.org/globalassets/global/toolbox/references/the-environmental-crisis.-the-environments-role-in-averting-future-food-crises-unep-2009.pdf. Accessed 2 May 2018.

National Oceanic and Atmospheric Administration (NOAA), 2015. Gulf of Mexico dead zone 'above average'. US Department of Commerce. www.noaanews.noaa.gov/stories2015/080415-gulf-of-mexico-dead-zone-above-average.html. Accessed 2 May 2018.

National Oceanic and Atmospheric Association (NOAA), undated. The Causes of Hypoxia in the Northern Gulf of Mexico. http://service.ncddc.noaa.gov/rdn/www/media/documents/hypoxia/hypox_finalcauses.pdf. Accessed 2 May 2018.

Owen, J., 2005. Farming claims almost half Earth's land, News Maps Show. National Geographic News. http://news.nationalgeographic.com/news/2005/12/1209_051209_crops_map.html. Accessed 2 May 2018.

RSPB, 2012. 44 million birds lost since 1966. www.rspb.org.uk/news/details.aspx?id=329911. Accessed 2 May 2018.

Sauer, J.R., Hines, J.E., Fallon, J.E., Pardieck, K.L., Ziolkowski Jr, D.J. and Link, W.A., 2014. The North American breeding bird Survey, results and analysis 1966–2013. *USGS Patuxent Wildlife Research Center.* www.mbr-pwrc.usgs.gov/bbs/. Accessed 2 May 2018.

Secretariat of the Convention on Biological Diversity, 2014. *Global Biodiversity Outlook 4.* Montréal. Accessed at: www.cbd.int/gbo/gbo4/publication/gbo4-en.pdf. Accessed 2 May 2018.

Soyatech, undated. Growing opportunities, soybeans and oilseeds. www.soyatech.com/soy_oilseed_facts.htm.

Stewart, A., 2015. Soy frontier at middle age. *2.* Hertz farm management, Inc. www.hertz.ag/ag-industry/current-headlines/0702bf5305182015112700/

Syngenta, undated. Why is soil so important? www.syngenta.com/global/corporate/SiteCollectionImages/Content/news-center/full/2014/why-is-soil-so-important-syngenta-infographic.pdf.

Veterinary Medicines Directorate (VMD), 2013. UK Veterinary antibiotic resistance and sales surveillance. www.gov.uk/government/uploads/system/uploads/attachment_data/file/440744/VARSS.pdf. Accessed 2 May 2018.

Virginia Institute of Marine Science (VIMS), 2018a.Trends. Low-oxygen 'dead zones' are increasing around the world. www.vims.edu/research/topics/dead_zones/trends/index.php. Accessed 2 May 2018.

Virginia Institute of Marine Science (VIMS), 2018b. Dead zones, lack of oxygen a key stressor on marine ecosystems. www.vims.edu/research/topics/dead_zones/index.php. Accessed 2 May 2018.

Worm, B., Barbier, E.B., Beaumont, N., Duffy, J.E., Folke, C., Halpern, B.S., Jackson, J.B.C., Lotze, H.K., Micheli, F., Palumbi, S.R., Sala, E., Selkoe, K.A., Stachowicz, J.J. and Watson, R., 2006. Impacts of biodiversity loss on ocean ecosystem services. *Science, AAAS.* 314(5800) 787–790. http://web.stanford.edu/group/Palumbi/manuscripts/impacts%20of%20biodiversity%20loss%20on%20ocean%20ecosystem%20services.pdf. Accessed 2 May 2018.

WWF Global, 2014. The growth of soy, impacts and solutions—The market for soy in Europe. http://wwf.panda.org/what_we_do/footprint/agriculture/soy/soyreport/the_continuing_rise_of_soy/the_market_for_soy_in_europe/. Accessed 2 May 2018.

# 4 Biodiversity, extinction and livestock production

*Tony Juniper*

The rising tide of human demand has become grossly unsustainable. Many ecosystems and renewable natural resources are under growing pressure. If human wellbeing is to be secured into the future then change is urgently needed. Analysis reveals that one of the top priority areas for action is the food system.

## Population and resources

Among the large-scale drivers that have created the current situation is that of rapid population growth. During the early 19th century the world population was about one billion people. In the late 1920s it had doubled to reach two billion and then three billion in 1960. By 1998 the population had doubled again, to reach six billion. In 2011 it passed seven billion, and in April 2017 it went over seven and a half billion. The number of humans is increasing annually at about the equivalent of the population of Germany and is expected to be well over nine billion by the middle of the present century (Figure 4.1).

This rise in the number of people has increased the demand for resources but not as much as the rapid increase in the size of the global economy. For a while, between 1950 and 2010 the population increased by about three times (United Nations Secretariat 2013), the economy expanded about tenfold (De Long 1998, World Bank website). People on average thus became richer, leading to the expansion of the global middle class, a trend that is expected to continue to the point where in the 2030s the majority of middle-class consumers are expected to be Asian, a dramatic change from the late 20th century, when most of these high consumers lived in North America and Europe (Figure 4.2).

## Urbanisation and diets

Economic growth has gone hand in hand with rapid urbanisation. A landmark moment was reached in relation to this trend when in 2007 for the first time in history more than half of the world's population lived in towns and cities (United Nations Secretariat, 2014). Up until that point the majority

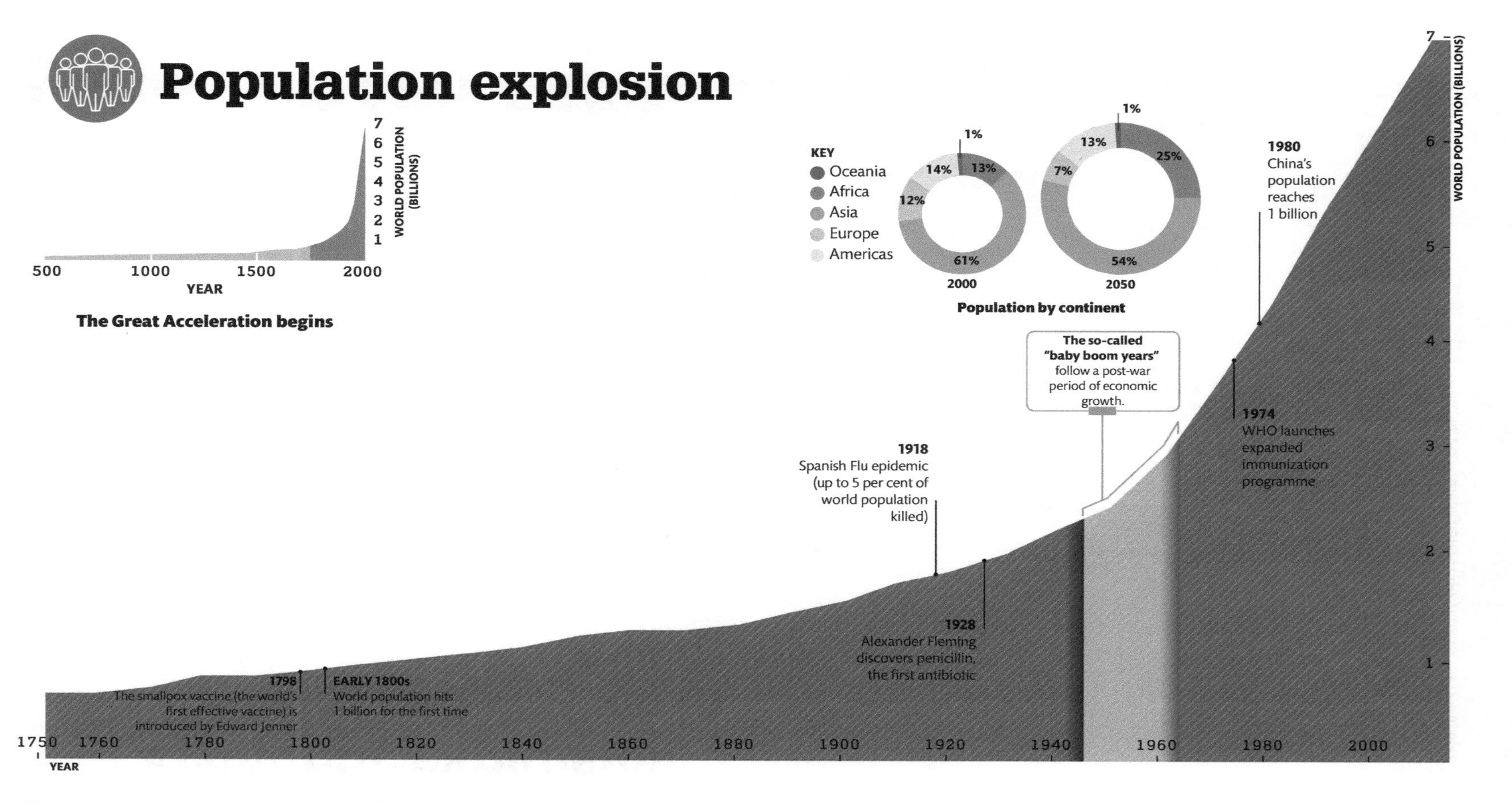


*Figure 4.1* Population explosion © 2018 Dorling Kindersley Limited.

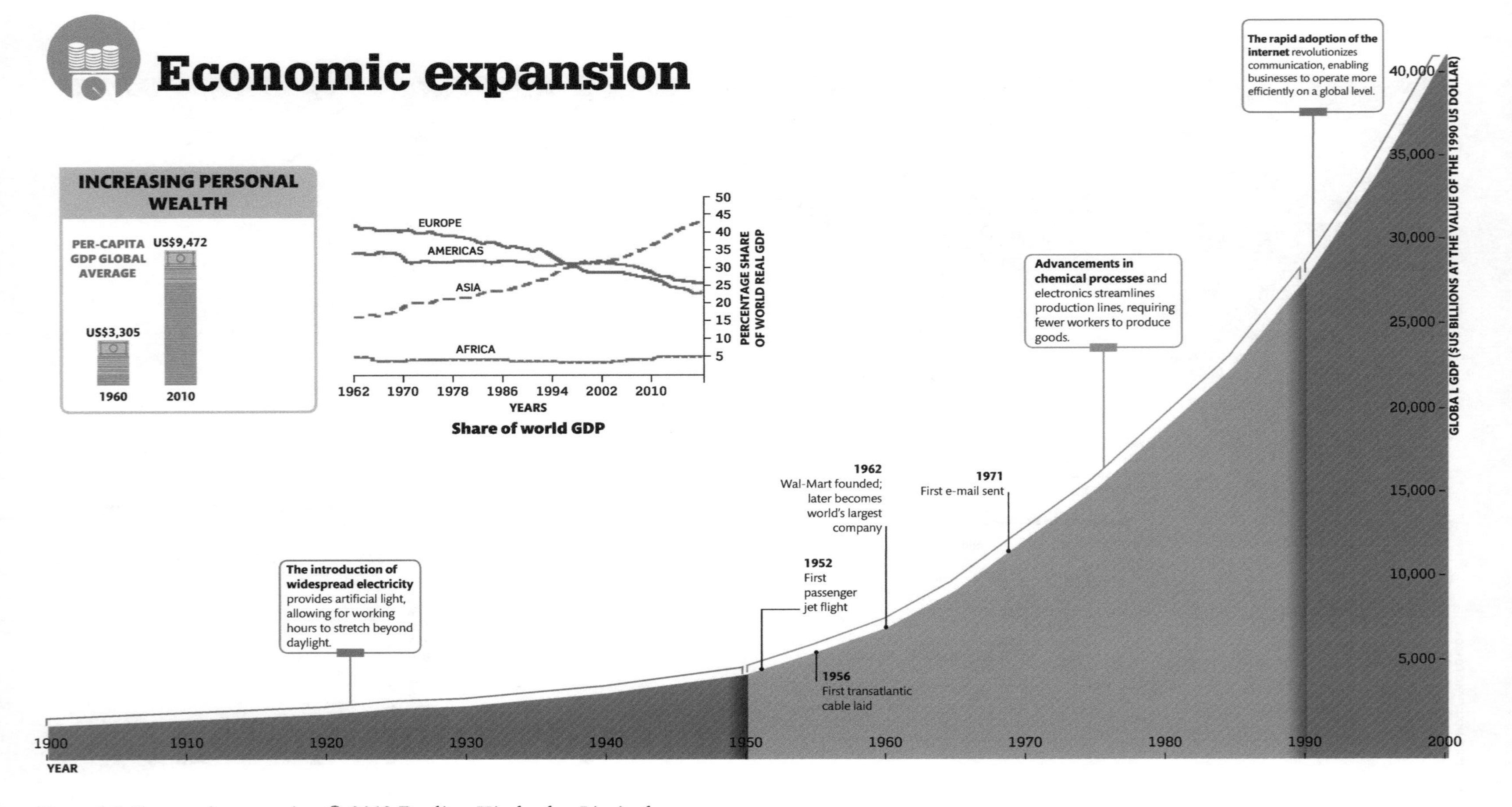


*Figure 4.2* Economic expansion © 2018 Dorling Kindersley Limited.

of us lived in rural environments, many occupied with farming. The rising urban middle class is expected to continue to grow so that by mid-century between two-thirds and three-quarters of us will live in urban areas, and by then the population will be far larger. While cities occupy only about 2% of the land, they are responsible for about three-quarters of the environmental impact that is being caused by human demand. This is in some large part linked with the supply of food and energy needed to sustain the fast-expanding city-dwelling population.

Richer consumers living in cities tend to have different diets to poorer people. With more money they have more choice and for many of them better diets has meant an increase in the consumption of livestock products. This is one reason why the consumption of meat and dairy products has during recent years run ahead of rising population. Far more energy, water and land are needed to sustain diets rich in livestock compared with those mainly comprised of plants, and so the rise of meat eating is another multiplier of impact that sits on top of rising human numbers. Those impacts of rising demand are now clear for all to see.

## Carbon dioxide and the food system

For example, the combustion of fossil energy is the main reason why carbon dioxide concentrations in the atmosphere have recently increased (IPCC, 2007). In 2013, and for the first time in about two million years, the level of this greenhouse gas went above 400 parts per million (US Department of Commerce, 2013 & International Geosphere-Biosphere Programme, 2015). Other factors have also contributed to this situation, including deforestation and soil damage. Both of those sources of carbon dioxide are primarily driven by our food system, including the clearance of forests to make way for soya to feed to captive pigs, chickens and cattle. Additional greenhouse gas emissions in the form of methane and nitrous oxide are also mainly derived from the food system, including the rearing of livestock (Figure 4.3).

## Habitats and ecosystems

More profound even than the changes human activities have caused to the atmosphere is how we have transformed terrestrial ecosystems. Demand for land for agriculture, including that growing crops to feed animals in factory farms, has led to massive-scale ecosystem damage and is the main cause of the loss of natural habitats. One measure of the scale of overwhelming impact on the biosphere from human activity is the estimate that whereas 10,000 years ago about 99.9% of terrestrial vertebrate biomass was comprised of wild animals, today about 96% of terrestrial vertebrate biomass is comprised of humans and their domesticated animals (Peak Oil Barrel). A high proportion of those captive animals dwell in factory farms, where they are reared on crops grown with impacts on habitats, water, energy and soils. In the oceans too

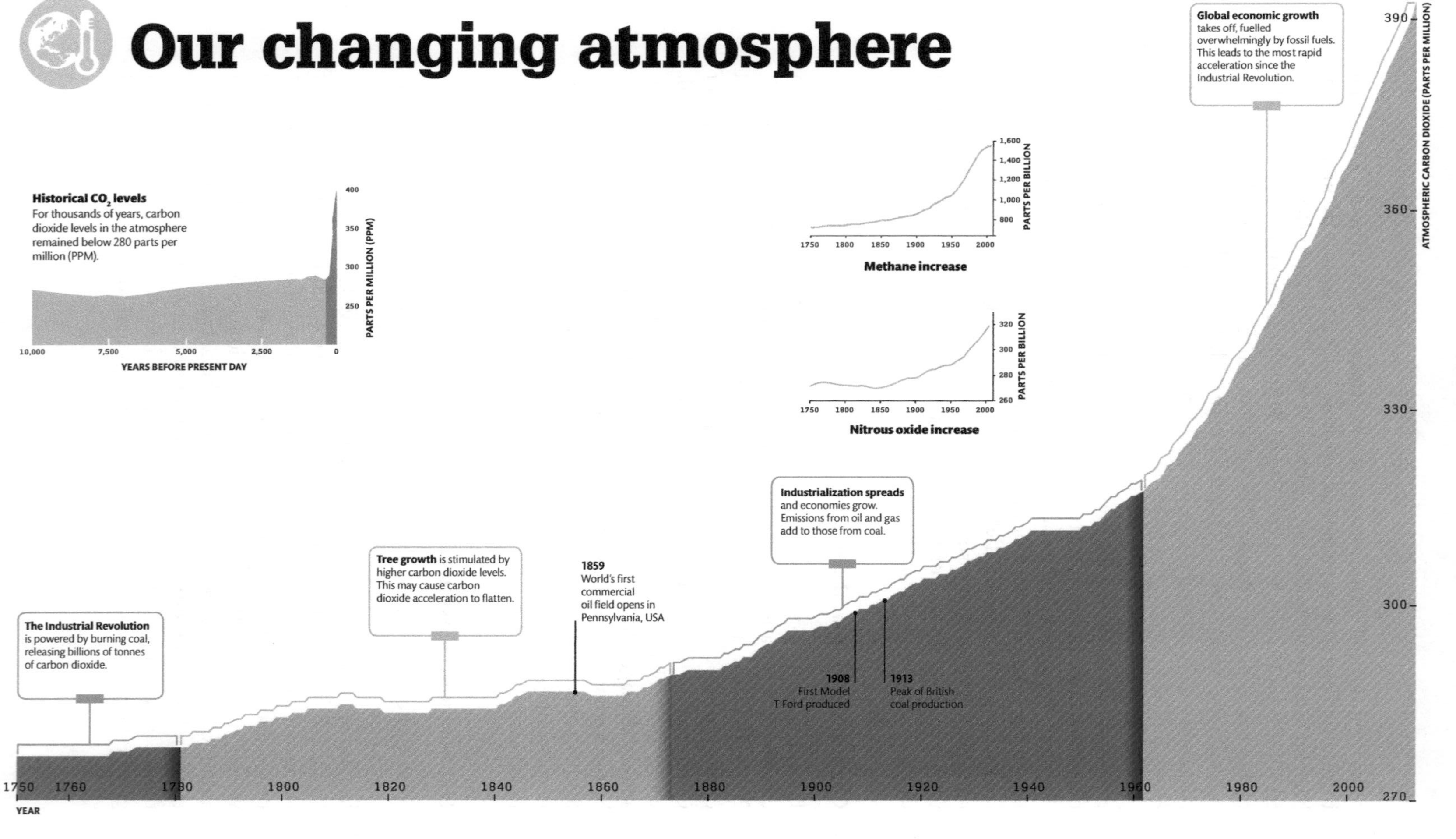

*Figure 4.3* Our changing atmosphere © 2018 Dorling Kindersley Limited.

several serious changes are taking place because of land-based factory farming of animals. These include coastal pollution and the depletion of some fish populations to feed captive animals, including farmed fish.

## Extinction

Behind all the manifestation of rising impact, in the end driven by population and economic growth translating into expanding demand for resources, is perhaps the most troubling consequence of all – namely the mass extinction of species. Now running between 100 and 1,000 times faster than background rates of species loss (Ceballos, G. et al 2015), a catastrophic decline in natural diversity is taking place, the largest-scale such event since the extinction of the dinosaurs about 65 million years ago. At the centre of this ecological crisis is our food system (Figure 4.4).

## Reforming our food and farming systems

That system is not only a cause of decline but also longer term will be a casualty of the dramatic changes taking place, as ecosystem services that underpin food production are progressively depleted, including the populations of pollinating animals, soil health and natural features that sustain water security. It is therefore the case that deep change is required in how we feed our expanding population, both for reasons of environmental sustainability and food security.

This, in turn, must depend upon a more rational assessment of the benefits and impacts of different food system choices, moving beyond the now outdated frames of cheapness and abundance that have dominated policy. These politically expedient and easy to understand objectives of food policy have hitherto trumped all other concerns, including the very considerable ecological costs that have accompanied industrialised and factory farming. If we are to meet different environmental targets, including in relation to climate change and biological diversity, then this must change.

The new frame for food and farming policy must be sustainability, not only for the sake of climatic stability and disappearing species but also for the very security of the nutrition needed to sustain our still expanding population. By moving towards a more rational food system, the benefits would not only be seen in the health of the environment and resilience of the system but also in the diminished suffering of farm animals and improved human health indicators.

In addition to more intelligent economic calculations as to the wisdom of our present direction of travel, there is an urgent need to shift research and policy signals in support of agroecological methods, including a reassessment of the place of livestock products in our collective diet. Joining up policy agendas, so that, for example, environmental, health, farming and rural development policies more closely align in pursuit of long-term broad-scale

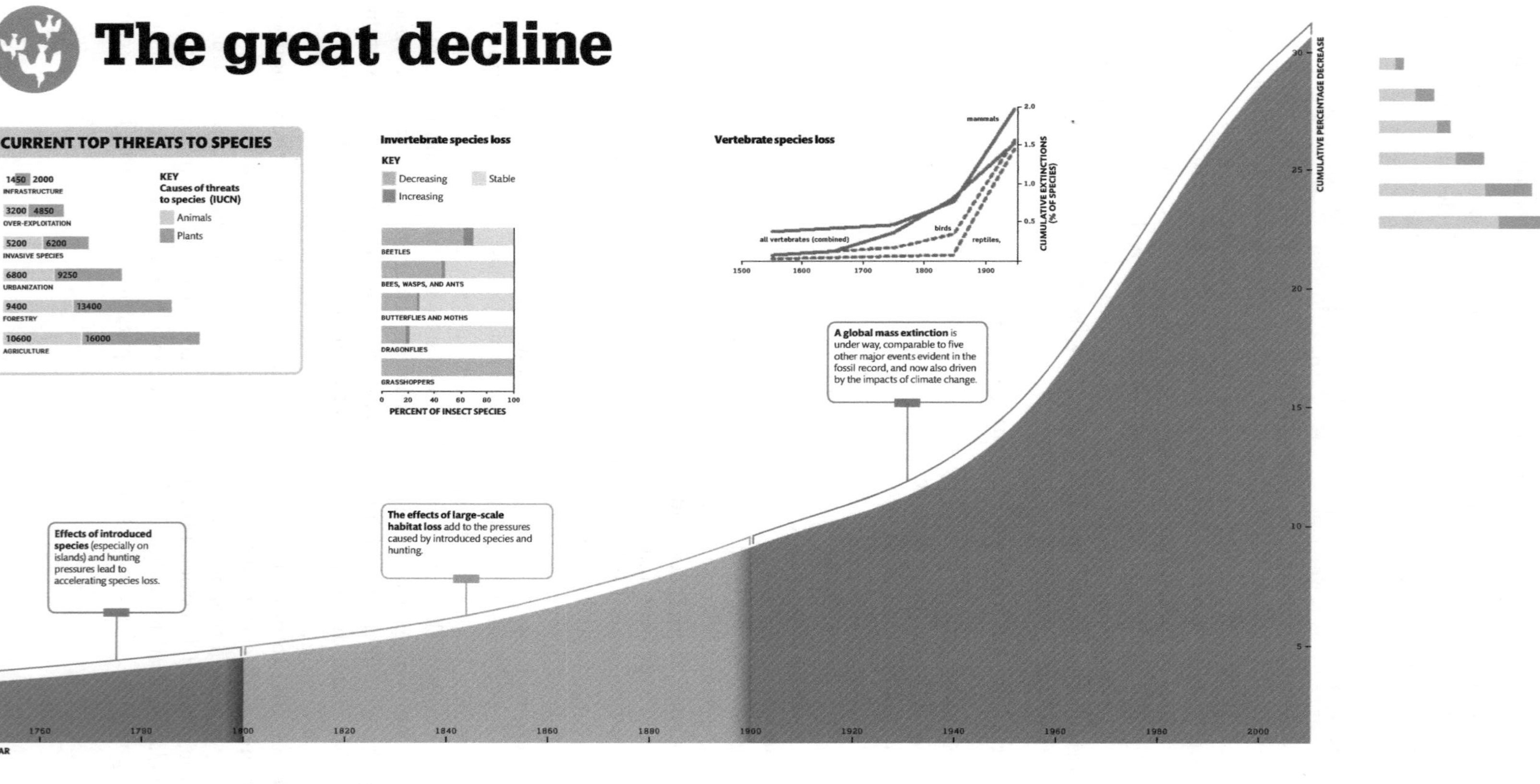


*Figure 4.4* The great decline © 2018 Dorling Kindersley Limited.

objectives is essential. This can be done but will require leaders and opinion formers to move beyond the outdated frames of reference that have brought us to the dangerous position we find ourselves in now.

The industrialised model of food production is no longer fit for purpose, including the manner in which we produce livestock products. It is literally unsustainable. The big question is whether it will end through a carefully thought through transition to a new system, or if it will collapse.

Further information can be found in Tony Juniper's 2016 book "What's Really Happening to our Planet?" (Published by DK Publishing).

## References

Ceballos, G., Ehrlich, P., Barnosky, A., García, A., Pringle, R., and Palmer, T. 2015. Accelerated modern human-induced species losses: Entering the sixth mass extinction. Science Advances. Accessed at http://advances.sciencemag.org/content/1/5/e1400253.full

De Long, World Bank. Compiled from research by J. Bradford De Long, Department of Economics, UC Berkeley (1998) and World Bank GDP data. Accessed at http://data.worldbank.org/indicator/NY.GDP.MKTP.CD, adapted to 1990 USD dollar prices.

International Geosphere-Biosphere Programme, 2015. 'Great Acceleration'. Accessed at www.igbp.net/globalchange/greatacceleration.4.1b8ae20512db692f2a680001630.html

IPCC, 2007. 'The Carbon Cycle and the Climate System', IPPC fourth assessment report: Climate Change 2007. Accessed at //www.ipcc.ch/publications_and_data/ar4/wg1/en/ch7s7-3.html#7-3-1

Peak Oil Barrel. Based on information from 'Fossil Fuels and Human Destiny', Peak Oil Barrel. Accessed at http://peakoilbarrel.com/fossil-fuels-human-destiny/

United Nations Secretariat, 2013. Population division of the department of economic and social affairs. World population prospects: The 2012 Revision. New York: United Nations.

United Nations Secretariat 2014. Population division of the department of economic and social affairs. UN World Urbanisation prospects: The 2014 Revision. New York: United Nations. Accessed at: http://esa.un.org/unpd/wup/Publications/Files/WUP2014-Highlights.pdf

US Department of Commerce, 2013. National Oceanic & Atmospheric Administration (2013), 'CO2 at NOAA's mauna loa observatory reaches new milestone: Tops 400 ppm'. Accessed at www.esrl.noaa.gov/gmd/news/7074.html

# 5 Bees versus robots

*Dave Goulson*

The decline of bees has attracted media headlines and widespread public concern in recent decades, and rightly so. Three-quarters of all the crops we grow benefit from insect pollination; many would produce little or nothing without pollinators (Gallai et al. 2009). Imagine a world without tomatoes, strawberries, coffee or chocolate to name just a few. Hence humankind should be deeply troubled by the ongoing declines in wild bee populations and by the rising mortality of domestic honeybee colonies. Some bee species are now extinct; for example Franklin's bumblebee, a native of the US, has not been seen since 2006 (Goulson et al. 2015). Farming needs bees, but paradoxically it is wiping them out; industrial farming involving large monocultures of crops treated with perhaps 20 different pesticides per year has made vast tracts of the globe into a hostile environment for wildlife of all sorts.

However, we may not have to worry about the bees for much longer; help is at hand. They could soon be redundant for there are plans to replace them – with robots. Teams of scientists from far-flung places, from Japan to Indonesia to the UK and US, are working on it as I write. There have been a number of scientific papers published discussing the possibility of building miniature flying robots to replace bees and pollinate our crops for us (Wood et al. 2013; Amador & Hu 2017; Barnett et al. 2017). Clumsy robobee prototypes have been built, and some seem to crudely work, although most still rely on a human to control them from a remote handset, and some seem more likely to chop flowers to pieces with their tiny rotor blades than to pollinate them. Regardless of these shortcomings, media coverage has heralded the imminent retirement of "the bee" and a brave new world in which tiny metal and plastic drones buzz from flower to flower. If crops could be pollinated this way, farmers wouldn't have to worry about harming bees with their insecticides. With wild bee populations in decline, perhaps these tiny automatons are the answer?

While I can understand the intellectual interest and challenge for a robotics engineer of trying to create robotic bees, I would argue that it is exceedingly unlikely that we could ever produce something as cheap (i.e. free) or as effective as bees themselves. Bees have been around and pollinating flowers for more than 100 million years; they have become exceedingly good at it. It is remarkable hubris to think that we can replicate or improve on them.

Consider just the numbers; there are roughly 80 million honeybee hives in the world, each containing perhaps 40,000 bees through the spring and summer (Aizen & Harder 2009). That adds up to 3.2 trillion bees, give or take. They feed themselves for free, breed for free and even give us honey as a bonus. What would be the cost of replacing them with robots? Even if the robots could be built, complete with charged power pack and control devices, for one penny each (which seems absurdly optimistic) it would cost £32 billion to build them. And how long would they last? Some would malfunction, some would get caught out in the rain or get lost, some would be damaged by wind or spiders' webs or curious bee-eaters. If we very optimistically calculate the lifespan of a robot bee at one year, that means spending £32 billion every year (and continually littering the environment with trillions of tiny robots, unless they could be made biodegradable). What about the environmental costs of manufacture and distribution? What resources would they require, what carbon footprint would they have, what energy source would power them? What would happen when terrorists or the Russians hacked into the robobee control system and turned them against us? Real bees avoid all of these issues; they are self-replicating, self-powering, essentially carbon neutral and unlikely to be subject to mind control by Vladimir Putin any time soon.

Thus far I have glossed over a vital further point. Pollination is not all done by honeybees. Numerous other insects pollinate crops and wild flowers, including butterflies, beetles, moths, flies, wasps, sawflies and many more. In more exotic climes hummingbirds, parrots and bats help out, and even occasionally lizards and marsupial mice. These pollinators come in all sorts of different shapes and sizes suited to different flowers. It has been calculated that honeybees contribute to at best one-third of crop pollination in the UK, averaged across crops, and that we have in the region of 4,000 other species of pollinator (Breeze et al. 2011). So, we wouldn't just need to replace the 3.2 trillion honeybees. We'd also need to replace countless trillions of other pollinators. All to substitute creatures that currently deliver pollination for free.

Declines of bees are symptomatic of larger issues. It is not just bees that are declining; almost all wildlife is declining in the face of massive habitat loss and pollution across the globe. Even supposing we could create robot bees cheaply enough for it to be viable, should we? If farmers no longer needed to worry about harming bees they could perhaps spray more pesticides, but there are many other beneficial creatures that live in farmland that would then be harmed: ladybirds, hoverflies and wasps that attack crop pests; worms, dung beetles and millipedes that help recycle nutrients and keep the soil healthy; and many more. Are we going to make robotic worms and ladybirds too? What kind of world would we end up with?

Do we have to always look for a technical solution to the problems that we create, when a simple, natural solution is staring us in the face? We have wonderfully efficient pollinators already, let's look after them, not plan for their demise.

This is adapted from my forthcoming book "The Garden Jungle".

## References

Aizen, M. & Harder, L.D. 2009. The global stock of domesticated honeybees is growing slower than agricultural demand for pollination. Current Biology 19: 915–918.

Amador, G.J. & Hu, D.L. 2017. Sticky solution provides grip for the first robotic pollinators. Chem 2: 162–164.

Barnett, J., Seabright, M., Williams, H., Nejati, M., Scarfe, A., Bell, J., … Duke, M. 2017. Robotic Pollination—Targeting kiwifruit flowers for commercial application. Presented at the PA17 International Tri-Conference for Precision Agriculture, Hamilton, New Zealand.

Breeze, T., Bailey, A.O., Balcombe, K.G., Potts, S.G. 2011. Pollination services in the UK: how important are honeybees? Agriculture, Ecosystems & Environment 142: 137–143.

Gallai, N., Salles, J-M., Vaissiere, B.E. 2009. Economic valuation of the vulnerability of world agriculture confronted with pollinator decline. Ecological Economics 68: 810–821.

Goulson, D., Nicholls E., Botías C., & Rotheray, E.L. 2015. Combined stress from parasites, pesticides and lack of flowers drives bee declines. Science 347: 1255957.

Wood, R., Nagpal, R., & Wei, G.-Y. 2013. Flight of the Robobees. Scientific American 308: 60–65.

# 6 Conservation and the sacred cow

*Dominic Wormell*

Working in conservation, you have to try to be optimistic. We are often battling against enormous forces to change natural habitats for the benefit of humans, frequently for agricultural use. But being so focussed on finding ways to save species and habitats, we can be guilty of not seeing the wood for the trees.

Some 30 years ago, when I started working in conservation, we thought both the causes and solutions to the conservation dilemma were complex, but over the years it has gradually dawned on me that the main cause for habitat destruction, here in the UK and internationally, was all around us: namely, our insatiable appetite for meat. So, I was keen to attend the 2017 Extinction and Livestock conference as it was the first time an international spotlight was being shone on the real elephant in the room: meat, the big issue of our times that affects the planet and our lives in so many profound ways.

For many years, I have focussed on the conservation of South American primates, and in particular, the marmosets and tamarins – the world's smallest monkeys. When I first went to Brazil's Atlantic rainforests to see rare marmosets, I didn't really think about cattle pasture as a threat. I generally thought, as I am sure most do, that human population growth, and the necessary expansion of agriculture, was simply how it was, and this was what we had left – tiny fragments of forest.

But every year for the last 20 years I have been travelling to South America, and in that time, I have seen huge changes. Cattle ranching is big business in Brazil, which is one of the largest exporters of beef in the world (Sharma, S. and Schlesinger, S., 2017). The vast cattle ranches are often owned by big international companies, with pretty much none of the profits going to local people.

The Atlantic forest biome, a "hotspot" with very high levels of endemism, once ran all the way down the eastern coast of Brazil. It has been reduced to just 7% of its former range. Now what you see when you visit this region is a vast expanse of open ground, often baked hard, with white zebu cattle as far as the eye can see or perhaps huge fields of sugar cane, most of which is grown for biofuel. Dotted around this virtual desert are isolated clumps of forest. But these are jewels of biodiversity that must not be lost.

The wonderful black lion tamarin, for example, is found only in the Brazilian state of São Paulo. It used to range right throughout that state but is now restricted to a few patches of isolated forest. There are probably only about 1,300 animals left. They are effectively trapped inside remaining forest fragments, surrounded by thousands of acres of cattle pasture and sugar cane plantations. The smallest of these forest islands may have as few as five individuals (Figure 6.1).

The black lion tamarin is by no means alone in suffering from the great loss of this unique and incredible rainforest system: the jaguars in the region are down to approximately 10–15 individuals. Many other species face similarly dire threats to their existence.

Such small populations are doomed to go extinct, as they are unable to move between the remaining patches of suitable habitat to find unrelated partners or establish new territories. Furthermore, such tiny remnants of forest are extremely vulnerable to stochastic events such as fires. Protecting and trying to reconnect these patches has become the conservation focus for those working to save the species. We will need to intervene directly if these fragmented wild populations of tamarins are to survive. The reintroduction and translocation of individuals between wild and captive populations will also be necessary to keep genetic flow within the species. In 1999 I took black lion tamarins, bred in captivity at Durrell Wildlife Conservation Trust, back to the species' main stronghold of Morro do Diabo in the first-ever reintroduction attempt for this species.

*Figure 6.1* The black lion tamarin: there are probably only 1,300 animals left. Photo: Dominic Wormell.

I have also been working with partners in Brazil to help join some of these isolated areas by growing tree corridors, and although we have made a great start, this is something that will take decades to achieve.

Through these intensive management efforts, we have stopped the black lion tamarin population from declining further, but it cannot increase in numbers because the remaining suitable habitat is so restricted.

One of the things that I have been struck by is that nobody talks about the issue that is driving these species to extinction. Several years ago, I was at a big lion tamarin conservation meeting that was attended by many people from around the world, who worked with these species. The astonishing thing was that every night we would go out to a steak house, with pretty much all the delegates tucking in. What if a group of orangutan conservationists went to a palm oil restaurant every night? Effectively eating rainforest. Some of the conservationists I work with raise an eyebrow and look surprised when I say I don't eat meat.

Sadly, I have come across the same situation with other species of tamarin that I have been working with. The white-footed tamarin of Colombia, found only in the Magdalena valley, is threatened as huge areas of land have been cleared for agriculture. One of the main uses for the land is cattle pasture (Boucher, 2011). Back in Brazil, the pied tamarin, found in the heart of the Amazon in and around the city of Manaus, is considered to be the most threatened primate in the Amazon basin. The city is expanding rapidly, converting the forest habitat into built-up areas. The demand for beef has meant that surrounding land is being converted to cattle pasture, and the tamarin is now critically endangered (Figure 6.2).

*Figure 6.2* Cattle ranching destruction in the vicinity of the Morro do Diabo reserve. Photo: Dominic Wormell.

So, there is a common theme running through all my experience in trying to conserve endangered tamarins in South America: clearance of land for livestock production, either for grazing or for soy to provide livestock feed for intensive systems throughout the world. Some 30% of South America is under the hoof (Trading Economics, 2018), and over 70% of the forest cleared is now pasture (De Sy et al., 2015). Yet despite this, and recent research that has shown that livestock production is the biggest cause for biodiversity loss, as outlined in other chapters in this book, conservation non-governmental organisations (NGOs) are conspicuous in their silence!

Conservationists talk a lot about the devastation caused by palm oil and invasive species. I read article after article discussing the urgent need to address the alien species issue in countries such as New Zealand and Australia. Introduced animals like the red fox or the cane toad do cause big problems for local endemic species, but surely these are dwarfed by the enormous scale of livestock farming in Australia and New Zealand, where millions of square kilometres are devoted to pasture. There are about 11,800 dairy herds, over 4.6 million cows, in New Zealand (Dairy Companies of New Zealand, 2018a), and 95% of all NZ milk produced is exported (Dairy Companies of New Zealand, 2018b). Water courses are seriously polluted and local habitats dramatically altered and damaged.

Yet we hear nothing about this from conservationists. No one dares mention livestock! Surely, the cattle and the sheep that are wrecking the habitats of unique fauna are the alien species with the most impact. When it comes to the UK we see a similar scenario with the sheep farming dominating the landscape of the uplands.

It sometimes seems that conservationists are content to focus on conserving small areas – fragmented scraps of forest, islands and other similarly small areas – and do not look beyond onto a grand landscape scale. Perhaps it seems too large a problem to contemplate, but we need to bring back great swathes of forests and other habitats, making the environment robust once more. It is the best way to mitigate against climate change, preventing floods and droughts, sequestering carbon from the atmosphere, changing local climatic conditions, securing water supplies for locals and providing livelihoods from the diversity a healthy ecosystem offers. Climate change will bring more and more unpredictable and extreme weather, and we have to restore ecosystems to mitigate against the worst floods and droughts that will come with such degraded landscapes. All of this is possible if humanity simply ends its love affair with meat. This has to be an international imperative. Whether restoring the forest of Brazil or the uplands of the UK, it will have huge benefits for all in the decades ahead.

## References

Boucher, D. (2011). Cattle and pasture. In: Boucher, D., Elias, P., Lininger, K., May-Tobin, C., Roquemore, S. and Saxon, E. (eds). *The Root of the Problem: What's driving tropical deforestation today?* Cambridge, MA: Union of Concerned Scientists.

Available at: www.ucsusa.org/sites/default/files/legacy/assets/documents/global_warming/UCS_DriversofDeforestation_Chap5_Cattle_1.pdf

Dairy Companies of New Zealand (2018a). QuickStats about dairying – New Zealand. www.dairynz.co.nz/media/5788611/quickstats_new_zealand_web_2017.pdf (downloaded on 25 April 2018).

Dairy Companies of New Zealand (2018b). About the NZ dairy industry. www.dcanz.com/about-the-nz-dairy-industry/ (downloaded on 25 April 2018).

De Sy, V., Herold, M., Achard, F., Beuchle, R., Clevers, J. G. P. W., Lindquist, E., & Verchot, L. (2015). Land use patterns and related carbon losses following deforestation in South America. *Environmental Research Letters* 10: 124004. Available at: http://iopscience.iop.org/article/10.1088/1748-9326/10/12/124004/pdf

Sharma, S., and Schlesinger, S., The Rise of Big Meat: Brazil's Extractive Industry, Institute for Agriculture and Trade Policy, 2017. Available at: https://www.iatp.org/sites/default/files/2017-11/2017_11_30_RiseBigMeat_f.pdf

Trading Economics (2018). Latin America & Caribbean (developing only) – agricultural land (% of land area). https://tradingeconomics.com/latin-america-and-caribbean/agricultural-land-percent-of-land-area-wb-data.html (downloaded on 25 April 2018).

# 7 Protected Cerrado and sustainable diets

## Complementary pathways towards a more conscious appetite

*Jean François Timmers*

How to conciliate the urgency of protecting the world's last natural ecosystems with the need to feed a fast-growing and deeply changing humanity? Using the Cerrado as a benchmark for this huge challenge inspires discussion of some of its multiple aspects and exploration of some initial solutions.

Food production causes 70%–80% of all global deforestation and biodiversity loss; causes most freshwater overuse, waste and pollution; empties the oceans to ever greater depths; and is the source of 25%–30% of all greenhouse gas emissions. Our appetite is actually our largest single impact on the planet and directly threatens our food as every year brings new climatic events, water crises, crop pests and other accidental effects of the chronic disruption of key ecosystem services upon which the resilience of agriculture, aquaculture and fishing depends (Rockström et al. 2009, Kissinger et al. 2012, Foley 2015, Gladek et al. 2016, World Wildlife Fund 2016).

Worse still, food demand is increasing steadily and is expected to at least double by 2050, multiplying these impacts (Alexandratos and Bruinsma 2012, Moomaw et al. 2012, World Resources Institute 2014). Population is growing in regions where food security is already seriously at stake, and urban middle class is booming unprecedentedly (Kharas 2017). We, the middle class, are responsible for most of the consumption and impacts of food. Our diets, rich in animal protein and processed products, generate most of the pressure on natural habitats, water and other natural resources. And our number grows annually at a rate around ten times higher than that of the total world population. Our consumption grows even faster, at an average rate of 4% yearly. Global demand for animal products (meat, dairy and fish) have doubled since the 1950s and following the actual trend may quadruple by 2050 (Moomaw et al. 2012, Gladek et al. 2016).

In 2020 the Earth will have to sustain at least two times as many high impact urban consumers as compared to 2010, representing for the first time the majority of humanity. In 2028, the number of such consumers is expected to reach more than five billion. At the same time, a majority of the world's two billion farmers are poor, and almost one million people today are at least periodically hungry – including many farmers. These crises will deepen as global food demand rises, pressing on lands, resources and prices.

Massive shift of diets and consumption patterns may then be essential to ensure the sustainability and future of food (Moomaw et al. 2012, Gladek et al. 2016). Nevertheless, a global consumption shift involves changes to individual and collective habits, values, identities and traditions, and this will inevitably demand time – at least more than a generation. We can already see clear signals that this transition is starting, generally with urban youth in many countries. Unfortunately, at the actual rate of environmental destruction, our planet won't be able to wait.

So, we face a double challenge: shifting our production, trade and consumption models at a massive scale as soon and as quickly as possible while concomitantly buying ourselves enough time for this transition by drastically reducing the current impacts of our actual food production. Massive efforts must be mobilised immediately to stop the destruction of natural habitats and watersheds, the pollution of freshwater and the oceans, widespread losses of soils, depletion of other key resources and the emissions of greenhouse gases (GHGs), among many other issues. This may protect us from an eventual global environmental collapse, maintaining and even restoring the key ecosystem services needed to support both our actual and, hopefully, our future more sustainable food systems.

Our fate may seem not so dramatic as half of the original ecosystems are still in existence and apparently available for expanding our food production (Ellis et al. 2013). It is unfortunately not that simple: most are frozen or deserts. Remaining spaces for agricultural expansion are essentially tropical and subtropical rainforests and savannahs. This is where biodiversity is at its highest, where most of the remaining wildlife may still be encountered and where deforestation and conversion of other natural habitats are actually most intense (Olson et al. 2001, Millennium Ecosystem Assessment 2005 p. 4, Barthlott et al. 2007, FAO 2010 p. 18, Kissinger et al. 2012, European Commission 2013 p. 45).

A recent study from World Wildlife Fund (WWF) identified that more than 80% of future deforestation is likely to be concentrated in 11 places, of which 10 are in the tropical and subtropical regions of South America, Africa and South-East Asia (World Wildlife Fund 2015).

Greater awareness and public concern have resulted in many global companies, governments and platforms issuing public commitments, since 2010, to eliminate by 2020 all deforestation from production, commodity-sourcing and financing. This was an unprecedented mobilisation and a very promising step towards achieving deforestation-free supply chains for food, fibre and energy. The Consumer Goods Forum (CGF), the New York Declaration on Forests, the Banking Environmental Initiative (BEI)(Soft Commodities Compact), Tropical Forest Alliance 2020 (TFA-2020), the Amsterdam Declarations and the United Nations (UN) Sustainable Development Goals (Target 15-2) all pledge to eliminate deforestation, mostly associated with the expansion of agriculture by the end of this decade.

As this deadline is approaching fast, there are growing concerns about achieving results, as progress is slow on the ground, and global deforestation is still steeply increasing (World Resources Institute 2017).

## Cerrado: the world's largest frontier and our best short-term opportunity

Oceans, lakes and rivers apart, global pressure from food production goes far beyond forests. The expansion of agriculture also threatens large pristine regions of savannahs, scrublands, wetlands and natural grasslands in the tropics, and also in North America and Central Asia. These ecosystems sometimes mix with forests and with each other in rich gradients and mosaics of interdependent ecosystems.

The impact of their destruction on nature and biodiversity is as overwhelming as deforestation. Their unique biological richness, their role in freshwater, protection of soils as carbon stock and protection against natural hazards, is now slowly being recognised. As this awareness grows, so is the use of the term "conversion of natural habitats", to designate the deep degradation or total replacement of any natural ecosystem by other land uses. Deforestation is the conversion of natural forests, among many other invaluable natural habitats.

The largest and most active deforestation/conversion front on the planet today is that of the Cerrado. The Cerrado is not as well known as the Amazon. It is nevertheless the world's most ancient and richest savannah, representing 5% of all the world's plant and animal species, including 4,800 endemic species of plants and vertebrates[1] (Critical Ecosystem Partnership Fund 2017, Prager and Milhorance 2018a). With its almost 2,000 tree species, the Cerrado is far richer than many of the world's forests.

This richness has been destroyed at an average rate of one million hectares' loss per year for the last 15 years to give space to the expansion of planted grazing pastures (of African grasses) for cattle and commercial crops – mainly soy (de Sy et al. 2015). The scale of this conversion was even larger in the previous decade (Brazilian Ministry of the Environment 2017). The actual rate of conversion is nevertheless bigger than that of the Amazon, which loses on average 600,000 hectares per year, while the Cerrado's total size is half of the Amazon's. This makes the deforestation/conversion impact on the Cerrado proportionally at least 2.5 times more intense. It may soon lead to 480 endemic plant species becoming extinct, which would represent over three times all documented plant extinctions since the European arrival in Brazil in the year 1500 (Strassburg et al. 2017).

To this day, 16% of the Amazon has been converted, and half of all the Cerrado has been annihilated, while another 30% suffered some level of degradation but may mostly recover spontaneously. The remaining 20% of the Cerrado is intact still, although less than 3% is secured in protected areas with strong restrictions to land use and conversion (Critical Ecosystem Partnership Fund 2017, Strassburg et al. 2017).

The Cerrado is a key watershed, pouring water into eight of the largest Brazilian river basins, essential for the future of its agriculture and most of the country's largest and most populated cities. In two decades it has become one of the world's largest and most strategic agricultural regions. And we now know that the vegetation of the Amazon and the Cerrado generates a significant part of their own rainfall, sending water in the form of clouds down to Argentina and Paraguay (Sheil and Murdiyarso 2009, Nobre 2014, Wright et al. 2017). Correlations between the deforestation of the Southern Amazon transition to the Cerrado and crop failures were detected. Currently additional evidence is being collected on the relationships between habitat destruction, climate anomalies and yields. Models to forecast impacts are elaborated (Input, unpublished data 2017). The Cerrado's conversion to agriculture is possibly directly threatening the future of the Cerrado's agriculture itself.

So, the challenge is to drastically reduce this world record front of destruction while meeting the expected booming demand for soy and beef in the next decades. And, apparently, this may be possible in just a few years, for the following three reasons.

First, it is technically feasible; there is enough available cleared land, highly suitable for soy, to double or even triple the actual soy production without cutting down one more single Cerrado tree or bush. This fact is now confirmed by at least three distinct research studies, and the suitable areas for conversion-free soy and cattle expansion are precisely mapped (Rudorff et al. 2015, Carneiro Filho et al. 2016, Rausch et al. in review). Also, a moderate increase in cattle ranching efficiency would free more than enough space for this conversion-free soy expansion, as well as sustainable cattle production (Strassburg et al. 2014). Actually this is already happening: in the last decade, 75% of soy expansion in the Cerrado occurred on already deforested, degraded pastures (Rudorff et al. 2015, Carneiro Filho et al. 2016), and most of the gains in Brazilian cattle production came from better practices more than from pasture expansion.

This shows that destruction of the Cerrado is neither necessary nor justifiable for food production, either for soy or cattle. Nevertheless, an additional key factor is land speculation. Cleared lands gain significant value, as compared to preserved Cerrado area, in the expectation of future use for soy (Steinweg et al. 2017, Prager and Milhorance 2018b). So, even when converted to pasture or simply abandoned after being deforested, soy is the prominent factor of lands' added value from deforestation or conversion. Strong consistent conversion-free soy demand and sourcing, the engagement of financial institutions and multi-stakeholder landscape planning have great potential to reduce this speculation bubble, efficiently and quite rapidly.

Second, we are experiencing unprecedented international awareness and mobilisation on the Cerrado, thanks to support from more than 60 global companies to the Cerrado Manifesto, and public acknowledgement from HRH Prince of Wales; at the same time, there is a strong growing signal from China, the largest and fastest growing soy and meat market, where 64 leading meat companies and associations signed the China Sustainable Meat Declaration, with an explicit pledge to avoid conversion of natural vegetation in the livestock and feed value chains.

Third, large-scale efforts are already on the way. Projects such as the Collaboration for Forests and Agriculture (CFA), funded by the Gordon and Betty Moore Foundation, and GEF Good Growth Partnership aim at increasing demand for deforestation/conversion-free soy and beef, and providing transparency tools for implementing the supply chain's commitments.

These three facts open a unique window of opportunity for decisive progress on curbing global deforestation and conversion. Eliminating conversion from soy and beef production in the Cerrado, and the Amazon, may curb as much as a third of all global actual and projected deforestation (World Wildlife Fund 2015). With an average daily Cerrado loss of more than 2.500 hectares, there is not one single day to lose.

The Cerrado can and must become a benchmark to curb or prevent other fast-growing deforestation/conversion fronts in the Chaco and other South American ecosystems as well as in the African Savannahs, Asiatic steppes and the last Northern American prairies.

## False dilemmas and right answers

Food is very personal to each human being, and discussions on the future of food may generate strong passionate debates. One dilemma often presented is the necessary choice between feeding a fast-growing humanity and protecting nature and tackling climate change. Of course, protecting nature is a pre-condition to feeding the world, as nature sustains all our food production, water, soils and biodiversity. More sustainable food production may be much more efficient, literally fixing carbon and helping soils to recuperate and improving freshwater quantity and quality.

There may also be no further justification for destroying any new ecosystems for more food production globally, considering the actual 40% food waste and loss, and the large-scale overconsumption of food.

Another common dilemma is whether to prioritise efforts to make food production more sustainable, to focus on reducing food waste and losses, or to shift diets to reduce food overconsumption and footprint. It seems clear that we need to do all three simultaneously. As discussed previously, it is essential to tackle the current impacts, gaining precious time for systemic changes, curbing the exponential growth of demand. All approaches are essential, interdependent and act on different time scales but must be tackled together to avoid the collapse of and ensure the future of the global food system.

A third dilemma is on the legitimacy of trying to secure natural ecosystems in regions still very rich in forests and savannahs, but with a booming population already in deep poverty and chronic or even critical food shortage. The destruction of ecosystems and related services usually deepen the crises, impact more directly on the poorest, and the related development often benefits new incoming populations. But it is possible to fight poverty while protecting the environment by using it more wisely and fighting fiercely the root causes of exclusion.

Innovative solutions are being designed and promising experiments are taking place. Many local projects have already demonstrated results on reducing poverty and protecting ecosystems, but still need to be transformed into broad policies. Daring strategies must be framed to tackle the global scale of the problem and be systemic as well as fast, as time is our rarest and most precious resource. Localised food chains may not prevail in total isolation. A simple embargo of all deforestation/conversion-related products and regions displaces the problem but does not solve it. Niche products will remain for the few and do not solve the global issues.

We need to tackle the impacts of food production matching the scale and pace of these impacts. In the Cerrado, which is the size of all Western Europe, this involves a broad massive international demand for conversion-free soy and beef, shrinking the land speculation bubble, together with multi-stakeholder pacts and public policies on where and how to use economically and to protect the available spaces more efficiently, and adequate technical and financial incentives to producers.

We only have a few years to try to contain all our food production within the geographical limits of what has already been destroyed, rehabilitating degraded soils, restoring watersheds and vegetation buffers. We also urgently need innovative solutions designed for regions where the last forests and savannahs remain and the populations are in critical need of development and support.

The Cerrado, the world's largest deforestation and conversion front, incarnates our global challenge. Let us save it, and start to save ourselves.

## Note

1 Total 11,430 plants (40% endemic), 837 birds, 199 reptiles, 251 amphibians and more than 1,300 fish species (Critical Ecosystem Partnership Fund, 2017).

## References

Alexandratos, N. and Bruinsma, J. 2012. World Agriculture Towards 2030/2050: the 2012 revision. ESA Working paper 12–03. Rome, FAO. Retrieved from www.fao.org/docrep/016/ap106e/ap106e.pdf.

Amsterdam Declaration 'Towards Eliminating Deforestation from Agricultural Commodity Chains with European Countries': Retrieved from www.euandgvc.nl/documents/publications/2015/december/7/declarations, accessed 09/06/2018.

Banking Environmental Initiative (Soft Commodities Compact): Retrieved from www.cisl.cam.ac.uk/business-action/sustainable-finance/banking-environment-initiative/programme/soft-commodities, accessed 09/06/2018.

Barthlott, W., Hostert, A., Kier, G., Küper, W., Kreft, H., Mutke, J., Rafiqpoor M. D. and Sommer J. H. 2007. Geographical patterns of vascular plant diversity at continental to global scales. *Erdkunde* 61(4): pp. 305–315. DOI: 10.3112/erdkunde.2007.04.01. Retrieved from https://www.researchgate.net/publication/215672847_Geographic_patterns_of_vascular_plant_diversity_at_continental_to_global_scale.

Brazilian Ministry of the Environment 2017. Os Planos de Prevenção e Controle do Desmatamento em Âmbito Federal (Deforestation Prevention and Control Plans at the Federal Level). Institutional website. http://combateaodesmatamento.mma.gov.br/, accessed 10/06/2018.

Carneiro Filho, A., Costa, K., Gesisky, J. et al. 2016. The expansion of Soybean production in the Cerrado, paths to sustainable territorial occupation, land use and production. agroicone, INPUT. São Paulo, Brazil. Retrieved from www.inputbrasil.org/wp-content/uploads/2016/11/The-expansion-of-soybean-production-in-the-Cerrado_Agroicone_INPUT.pdf.

Cerrado Manifesto: Retrieved from http://wwf.panda.org/wwf_news/press_releases/?310899/Environmentalists-ask-markets-to-help-stop-the-destruction-of-the-Cerrado, accessed 09/06/2018.

China Sustainable Meat Declaration: Retrieved from www.wwf.org.br/?61882/China-Meat-Association-And-Its-64-Chinese-Company-Members-Jointly-Announce-Chinese-Sustainable-Meat-Declaration-with-WWF, accessed 09/06/2018.

Collaboration for Forests and Agriculture (CFA): Retrieved from www.moore.org/docs/default-source/default-document-library/executive-summary_collaboration-for-forests-and-agriculture.pdf?sfvrsn=431c6d0c_0, accessed 09/06/2018.

Consumer Goods Forum Resolution on Deforestation: Retrieved from www.theconsumergoodsforum.com/initiatives/environmental-sustainability/key-projects/deforestation/, accessed 09/06/2018.

Critical Ecosystem Partnership Fund – CEPF 2017. Ecosystem profile of the Cerrado Biodiversity Hotspot. Revised version (Feb. 2017). Retrieved from www.cepf.net/our-work/biodiversity-hotspots/cerrado/.

de Sy, V., Herold, M., Achard, F., Beuchle, R., Clevers, J.G.P.W., Lindquist, E. and Verchot, L.V. (2015). Land use patterns and related carbon losses following deforestation in South America. *Environmental Research Letters* 10(12): p. 124004. DOI: 10.1088/1748-9326/10/12/124004. Retrieved from www.cifor.org/library/5892/land-use-patterns-and-related-carbon-losses-following-deforestation-in-south-america/.

Ellis, EC, Kaplan, JO, Fuller, DQ, Vavrus, S., Goldewijk, KK and Verburg, PH. 2013. Used planet: A global history. *Proceedings of the National Academy of Sciences of the United States of America* 110 (20): pp. 7978–7985. DOI: 10.1073/pnas.1217241110. Retrieved from www.pnas.org/content/110/20/7978.

European Commission 2013. The impact of EU consumption on deforestation: Comprehensive analysis of the impact EU consumption on deforestation. Technical Report -2013 – 063. European Union, Brussels. Retrieved from http://ec.europa.eu/environment/forests/pdf/1.%20Report%20analysis%20of%20impact.pdf.

FAO 2010. Global Forest Resources Assessment 2010. Main Report. FAO Forestry Paper 163. Food and Agriculture Organization of the United Nations. Rome. Retrieved from www.fao.org/docrep/013/i1757e/i1757e.pdf.

Foley, J. 2015. A Five Step Plan to Feed the World. National Geographic. Retrieved from www.nationalgeographic.com/foodfeatures/feeding-9-billion/, accessed 09/06/2018.

Gladek, E., Fraser, M., Roemers, G., Sabag Muñoz, O., Kennedy, E. and Hirsch, P. 2016. The global food system: An analysis. Metabolic, Amsterdam, The Netherlands. Retrieved from www.metabolic.nl/publications/global-food-system-analysis/.

Global Environment Facility – GEF Good Growth Partnership Project: Retrieved from www.undp.org/content/gcp/en/home/presscenter/articles/2017/08/29/major-players-get-ready-to-put-sustainability-at-the-heart-of-global-commodity-supply-chains.html, accessed 09/06/2018.

Kharas, H. 2017. The unprecedented expansion of the global middle class: An update. Brookings Global Economy & Development Working Paper 100. Retrieved from www.brookings.edu/wp-content/uploads/2017/02/global_20170228_global-middle-class.pdf.

Kissinger, G., Herold, M. and de Sy, V. 2012. Drivers of deforestation and forest degradation: A synthesis report for REDD+ Policymakers. Lexeme Consulting, Vancouver, Canada. Retrieved from https://assets.publishing.service.gov.uk/government/uploads/system/uploads/attachment_data/file/65505/6316-drivers-deforestation-report.pdf.

Millennium Ecosystem Assessment 2005. Ecosystems and human well-being: Synthesis. Island Press, Washington, DC. Retrieved from www.millenniumassessment.org/documents/document.356.aspx.pdf.

Moomaw, W., Griffin, T., Kurczak, K. and Lomax, J. (2012). The critical role of global food consumption patterns in achieving sustainable food systems and food for all, a UNEP discussion paper. United Nations Environment Programme, Division of Technology, Industry and Economics, Paris, France. Retrieved from http://wedocs.unep.org/bitstream/handle/20.500.11822/25186/Food_Consumption_Patterns.pdf?sequence=1&isAllowed=y.

New York Declaration on Forests (NYDF) Goal 2: Retrieved from http://forestdeclaration.org/goal/goal-2/, accessed 09/06/2018.

Nobre A.D. 2014. The future climate of Amazonia. Scientific Assessment Report. CCST-INPE, INPA, ARA. São José dos Campos, Brazil. Retrieved from www.ccst.inpe.br/wp-content/.../11/The_Future_Climate_of_Amazonia_Report.pdf.

Olson, D.M., Dinerstein, E., Wikramanayake, E.D., Burgess, N.D., Powell, G.V.N., Underwood, E.C., D'Amico, J.A., Itoua, I., Strand, H.E., Morrison, J.C., Loucks, C.J., Allnutt, T.F., Ricketts, T.H., Kura, Y., Lamoreux, J., Wettengel, W.W., Hedao, P., and Kassem, K.R. 2001. Terrestrial ecoregions of the world: New map of life on earth. *Bioscience* 51(11): pp. 933–938. Retrieved from www.researchgate.net/publication/216340317_Terrestrial_Ecoregions_of_the_World_A_New_Map_of_Life_on_Earth.

Prager, A. and Milhorance, F. 2018a. Cerrado: Appreciation grows for Brazil's savannah, even as it vanishes. Online News Article, 12 March 2018. Mongabay series: Cerrado. Retrieved from https://news.mongabay.com/2018/03/cerrado-appreciation-grows-for-brazils-savannah-even-as-it-vanishes/.

Prager, A. and Milhorance, F. 2018b. Cerrado: U.S. investment spurs land theft, deforestation in Brazil, say experts. Online News Article, 28 March 2018. Mongabay series: Cerrado. Retrieved from https://news.mongabay.com/2018/03/cerrado-u-s-investment-spurs-land-theft-deforestation-in-brazil-say-experts/.

Rausch, L.L., Gibbs, H.K., Schelly, I., Brandão Jr., A., Morton, D.C., Carneiro Filho, A., Strassburg, B., Walker, N., Noojipady, P., Barreto, P., Meyer, D. Soy Expansion in Brazil's Cerrado. Under review at Nature Sustainability.

Rockström, J., Steffen, W., Noone, K., Persson, Å., Chapin III, F.S., Lambin, E.F., Lenton, T.M., Scheffer, M., Folke, C., Schellnhuber, H.J. et al. 2009. A safe operating space for humanity. *Nature* 461(7263): pp. 472–475. DOI: 10.1038/461472a; pmid: 19779433. Retrieved from www.nature.com/articles/461472a.

Rudorff, B., Risso, J., Aguiar, D., Gonçalves, F., Salgado, M. et al. 2015. Geospatial analysis of the annual crops dynamic in the Brazilian Cerrado Biome: 2000 to 2014. Agrosatélite Applied Geotechnology Ltd. Florianópolis, Santa Catarina, Brazil. Retrieved from http://biomas.agrosatelite.com.br/img/Geospatial_analyses_of_the_annual_crops_dynamic_in_the_brazilian_Cerrado_biome.pdf.

Sheil, D. and Murdiyarso, D. (2009). How forests attract rain: An examination of a new hypothesis. BioScience 59(4): pp. 341–347. DOI: 10.1525/bio.2009.59.4.12. Retrieved from www.cifor.org/library/2770/how-forests-attract-rain-an-examination-of-a-new-hypothesis/.

Steinweg, T., Kuepper, B. and Thoumi, G. 2017. Farmland investments in Brazilian Cerrado: Financial, environmental and social risks. Chain Reaction Research. Aidenvironment, Climate Advisers, Profundo. September 20, 2017. Retrieved from https://chainreactionresearch.com/report/farmland-investments-in-brazilian-cerrado-financial-environmental-and-social-risks/.

Strassburg, B.B.N., Latawiec, A.E., Barioni, L.G., Nobre, C.A., da Silva, V.P., Valentim, J.F., Vianna, M. and Assad, E. D. 2014. When enough should be enough: Improving the use of current agricultural lands could meet production demands and spare natural habitats in Brazil *Global Environmental Change* 28(0): pp. 84–97. DOI: 10.1016/j.gloenvcha.2014.06.001. Retrieved from www.sciencedirect.com/science/article/pii/S0959378014001046.

Strassburg, B.B.N., Brooks, T., Feltran-Barbieri, R., Iribarrem, A., Crouzeilles, R., Loyola, R., Latawiec, A.E., Oliveira Filho, F.J., Scaramuzza, C.A.M., Scarano, F.R., Soares-Filho, B. and Balmford, A. 2017. Moment of truth for the Cerrado hotspot. *Nat Ecol Evol.* 1(4): p. 99. DOI: 10.1038/s41559-017-0099. Retrieved from www.nature.com/articles/s41559-017-0099.

Tropical Forest Alliance TFA-2020: Retrieved from www.tfa2020.org/en/, accessed 09/06/2018.

UN Sustainable Development Goals (Target 15-2): Retrieved from https://sustainabledevelopment.un.org/sdg15, accessed 09/06/2018.

World Resources Institute – WRI 2014. Creating a Sustainable Food Future. A menu of solutions to sustainably feed more than 9 billion people by 2050. World resources report 2013–14: interim findings. Retrieved from www.wri.org/sites/default/files/wri13_report_4c_wrr_online.pdf.

World Resources Institute – WRI 2017. Global tree cover loss rose 51 percent in 2016. Online article. Weisse M. and Goldman E. D. October 23, 2017. Retrieved from www.wri.org/blog/2017/10/global-tree-cover-loss-rose-51-percent-2016.

World Wildlife Fund – WWF 2015. Living forests report Chapter 5: Saving forests at risk. WWF International, Gland, Switzerland. Retrieved from www.worldwildlife.org/publications/living-forests-report-chapter-5-saving-forests-at-risk.

World Wildlife Fund – WWF 2016. Living Planet Report 2016. Risk and resilience in a new era. WWF International, Gland, Switzerland. Retrieved from wwf.panda.org/knowledge_hub/all_publications/lpr_2016/.

Wright, J. S., Fu, R., Worden, J. R., Chakraborty, S., Clinton, N. E., Risi, C., Sun, Y. and Yin, L. 2017. Rainforest-initiated Wet Season Onset over the Southern Amazon. *Proceedings of the National Academy of Sciences of the United States of America* 114(32): pp. 8481–8486. DOI: 10.1073/pnas.1621516114. Retrieved from www.pnas.org/content/114/32/8481.

# 8 The growing demand for fish

## Impacts on the environment, animal welfare and food security

*Krzysztof Wotjas and Natasha Boyland*

Global fish consumption has reached record levels, with the average person eating roughly 20 kg in 2014, an amount that has doubled since the 1960s and expected to rise further still (FAO, 2016). This demand is unsustainable and is having a heavy impact on ocean ecosystems, food security and fish welfare.

In 2014, over 93 million tonnes of fish were caught from the world's seas; a huge quantity, but one that has not increased much since the 1990s (FAO, 2016). Despite growing efforts – more boats, fuel and modern technology – fisheries are unable to catch more fish because they appear to have reached the ocean's upper limit. Ninety per cent of the world's commercially exploited fish populations are now being fully or over-fished (FAO, 2016). Some species, such as the southern Bluefin tuna, are threatened with extinction (World Ocean Review, 2013), and scientists are warning that if fishing activity continues at these levels there will eventually be no fish left in our oceans (Worm, 2016). Overfishing is currently driving the most rapid rates of evolution ever observed in wild populations (Palkovacs, 2011). There are also serious ramifications for non-fished species as biodiversity and ecological stability are affected by the huge numbers of fish being removed from the sea (Ellingsen et al., 2015). Fisheries cause substantial environmental damage. For example "bottom trawling", the practice of pulling a large fishing net along the bottom of the sea behind a ship, can remove around 5%–25% of the seabed life on a single run (Jackson at al., 2017).

While most discussions of commercial fishing are centred on sustainability and food security, animal welfare must not be overlooked. It is important to remember that the millions of tonnes of fish caught each year equate to 830 billion to 2.4 trillion individual sentient beings (estimates based on method from Mood and Brooke (2010), based on data from FAO Fishstat J 2015). Capture and landing methods therefore present huge welfare concerns. For example, catching fish in trawl nets involves guiding them to the mouth of the net then eventually exhausting and overrunning them as the net is towed forwards (Gregory, 1998). They are then pulled from the sea in large nets with thousands of others (Gregory, 1998). Many are killed during this process, but otherwise will be left to suffocate on deck or may be gutted alive (Van de Vis and Kestin, 1996).

Although fishery production may have peaked around 30 years ago, farming has expanded to meet the growing global demand for fish. Aquaculture is currently the fastest growing animal production sector, with output increasing from 5 to 63 million tonnes in 30 years, overtaking wild fisheries as the main source of fish for human consumption in 2014 (FAO, 2016; World Bank, 2013). The rapid growth of aquaculture is often called the "blue revolution" in reference to the "green revolution" that occurred between the 1930s and late 1960s, and resulted in increased agricultural production worldwide (Sachs, 2007). The key difference between the green and blue revolution, however, is that fish are not crops but sentient animals. Unfortunately, in most cases modern aquaculture resembles the factory farms that are all too commonly used on land for terrestrial farm animals.

Generally, aquaculture operations are set to be very intensive. Overcrowded fish are vulnerable to disease and experience stress, aggression and physical injuries such as fin damage (Person-Le Ruyet et al., 2008). Fish are crammed in sea cages or barren tanks with no consideration of their needs or natural behaviour. Halibut have a demersal lifestyle in the wild, a flatfish living on sandy and muddy sea floors. However, in fish farms, they are kept in cages with no substrate and it is common to see individuals trying to hide under each other in the absence of sand (Farm Animal Welfare Committee (FAWC), 2014). The Atlantic salmon is a migratory species that travels on average 5–30 km at sea per day and is solitary in its adult life (Willoughby, 1999). In fish farms they are kept in sea cages with hundreds or thousands of others, where these long-distance swimmers can travel only round the cage, and where parasites become endemic.

Salmon farming is also detrimental to the environment. High stocking densities and the unnatural life cycle cause severe problems with parasites and diseases (Ashley, 2007). Chemical treatments are often used to remove sea lice from salmon, which means releasing toxic agents into seawater (Burridge and Geest, 2014). Parasites have become increasingly resistant to the chemicals; so, alternative treatments are now being employed to attempt to control sea lice populations, but these come with their own welfare problems. One such alternative treatment is the Thermolicer – a machine which exposes salmon to water that is much warmer than they are accustomed to – which has led to the death of huge numbers of fish. For example, in March 2016, 32,700 salmon died on a farm in Norway during treatment for sea lice, of which 19,620 deaths occurred immediately after the treatment, with the remaining fish dying over the next week. The parasites that thrive in fish farms also attack wild fish, threatening local populations, and sea lice have become a global problem (Costello, 2009). Waste and antibiotics from sea cages can severely affect ecosystems (Friends of the Earth, 2018).

Like wild-caught fish, farmed fish are often killed with no respect to their welfare. Many die by suffocating in air or iced water, or exposure to high levels of carbon dioxide or they may be gutted alive (European Commission,

2017). But there are some higher welfare slaughter practices that involve stunning the fish before killing. Percussive and electrical stunning equipment is now available and becoming more applicable to different species. The aquaculture industry should urgently adopt humane slaughter methods for all species and types of production.

Although aquaculture is often presented as a solution to overfishing, it is currently increasing the pressure on wild fish populations. Some of the most popular farmed fish consumed in Europe – salmon, trout, sea bass and sea bream – are all highly carnivorous species (UNEP, 2009). Farming these species requires large amounts of fish meal and fish oil to feed them (Naylor et al., 2000). Those are obtained by grounding up wild-caught fish; up to half of all the fish caught in the wild are ground up to create feed for farmed animals, including fish and shrimps (estimates according to Mood, pers. Comms., 2018 based on data from FAO, 2015). This means that aquaculture often involves the suffering of hidden numbers of fish and is an inefficient use of resources.

At the same time, the evidence for fish sentience is growing, based on an accumulation of studies testing their brain structure, learning abilities and behaviour (Sneddon, Braithwaite and Gentle, 2003). Scientists are also documenting the complex behaviours of fish and demonstrating their impressive cognitive capacity (Brown, 2014). Researchers discover that some fish species take good care of their offspring (Goodwin et al., 1998) and can recognise faces (Siebeck et al., 2010), cooperate with other species (Bshary et al., 2006) and even use tools (Jones, Brown, and Gardner, 2011).

Fortunately, better aquaculture is possible. Farming herbivorous or omnivorous species that can be fed with plant-based feed can eliminate the destructive impact of fish farming on the ocean. Extensive and organic production is also possible under water (Yussefi, 2010). There are so many innovations in the way we produce our food. Innovations in the aquaculture sector should be focussed on improving fish welfare and limiting environmental impact rather than looking only at ways to maximise production. Industry efforts to replace fish meal and fish oil with plant-based substitutes or supplement it with fish trimmings from human consumption are examples of steps in the right direction (Carter and Hauler, 2000). There is, however, more effort needed. On the welfare side, we need to study and identify the natural needs of fish and adjust the farming systems for them. Lower stocking densities and environmental enrichment are part of the solution. We might also need to recognise that some fish species are not suitable for farming as the welfare and environmental cost is just too high.

We must also change the way we see fish. Consumers need to be aware of the broad range of issues and choose more consciously. Our urge to eat fish, and especially carnivorous species, is having disastrous effects. The conclusion that we need to limit our animal product consumption to save the planet is very apparent now. We must also remember that this needs to include fish.

## References

Ashley, P. J. (2007) Fish welfare: Current issues in aquaculture, Applied Animal Behaviour Science, 104(3–4), pp. 199–235. doi: 10.1016/j.applanim.2006.09.001.

Brown, C. (2014) Fish intelligence, sentience and ethics, Animal cognition. doi: 10.1007/s10071-014-0761-0.

Bshary, R., Hohner, A., Ait-el-Djoudi, K., & Fricke, H. (2006) Interspecific communicative and coordinated hunting between groupers and giant moray eels in the Red Sea. PLoS biology, 4(12), e431.

Burridge, L. E., and Geest, J. L. Van (2014) A review of potential environmental risks associated with the use of pesticides to treat Atlantic salmon against infestations of sea lice in Canada, DFO Can. Sci. Advis. Sec. Res. Doc. 2014/002, (March), p. vi + 36p.

Carter, C. G., & Hauler, R. C. (2000) Fish meal replacement by plant meals in extruded feeds for Atlantic salmon, Salmo salar L. Aquaculture, 185(3–4), pp. 299–311.

Costello, M. J. (2009) How sea lice from salmon farms may cause wild salmonid declines in Europe and North America and be a threat to fishes elsewhere, Proceedings of the Royal Society B-Biological Sciences, 276(1672), pp. 3385–3394. doi: 10.1098/rspb.2009.0771.

Ellingsen, K. E. et al. (2015) The role of a dominant predator in shaping biodiversity over space and time in a marine ecosystem, Journal of Animal Ecology, 84(5), pp. 1242–1252. doi: 10.1111/1365-2656.12396.

European Commission (2017) Welfare of farmed fish : Common practices during transport and at slaughter.

FAO (2016) The State of World Fisheries and Aquaculture. doi: 10.5860/CHOICE.50–5350.

Farm Animal Welfare Committee (FAWC) (2014) 'Opinion on the Welfare of Farmed Fish', (October).

FishStat, J. (2015) Fisheries and aquaculture software. FishStatJ-software for fishery statistical time series. FAO Fisheries and Aquaculture Department, Rome.

Friends of the Earth (2018) The Dangers of Industrial Ocean Fish Farming, pp. 1–27.

Goodwin, N. B., Balshine-Earn, S., & Reynolds, J. D. (1998) Evolutionary transitions in parental care in cichlid fish. Proceedings of the Royal Society of London B: Biological Sciences, 265(1412), pp. 2265–2272.

Gregory, N. G., and Grandin, T. (1998) Animal welfare and meat science (No. 636.08947 G7). CABI Pub.

Jackson W. J., Argent R. M., Bax N. J., Clark G. F., Coleman S., Cresswell I. D., Emmerson K. M., Evans K., Hibberd M. F., Johnston E. L., Keywood M. D., Klekociuk A., Mackay R., Metcalfe D., Murphy H., Rankin A., Smith D. C., and Wienecke B. (2017) Australia state of the environment 2016: overview, independent report to the Australian Government Minister for the Environment and Energy, Australian Government Department of the Environment and Energy, Canberra.

Jones, A. M., Brown, C. and Gardner, S. (2011) Tool use in the tuskfish Choerodon schoenleinii?, Coral Reefs, 30(3), p. 865. doi: 10.1007/s00338-011-0790-y.

Mood, A. and Brooke P. (2010) Worst things happen at sea: the welfare of wild-caught fish. Accessed 23 May 2013 at www.fishcount.org.uk.

Mood, A. (2018) Personal communication based on data from FAO, 2015.

Naylor, R. L. et al. (2000) Effect of aquaculture on wold fish supply, Nature, 405, pp. 1017–1024.

Palkovacs, E. P. (2011) The overfishing debate: An eco-evolutionary perspective, Trends in Ecology and Evolution, 26(12), pp. 616–617. doi: 10.1016/j.tree.2011.08.004.

Person-Le Ruyet, J., Labbé, L., Le Bayon, N., Sévère, A., Le Roux, A., Le Delliou, H., & Quéméner, L. (2008) Combined effects of water quality and stocking density on welfare and growth of rainbow trout (Oncorhynchus mykiss). Aquatic Living Resources, 21(2), pp. 185–195. doi:10.1051/alr:2008024.

Sachs, Jeffrey D. (2007) "The Promise of the Blue Revolution". Scientific American. Archived from the original on May 28, 2017. Retrieved 28 May 2017.

Siebeck, U. E., Parker, A. N., Sprenger, D., Mäthger, L. M., & Wallis, G. (2010) A species of reef fish that uses ultraviolet patterns for covert face recognition. Current Biology, 20(5), pp. 407–410.

Sneddon, L. U., Braithwaite, V. A., and Gentle, M. J. (2003) 'Do fishes have nociceptors? Evidence for the evolution of a vertebrate sensory system', Proceedings of the Royal Society B: Biological Sciences, 270(1520), pp. 1115–1121. doi: 10.1098/rspb.2003.2349.

United Nations Environment Programme (UNEP) (2009) The Role of Supply Chains in Addressing the Global Seafood Crisis.

Van de Vis H. and Kestin, (1996) Killing of fishes: literature study and practice-observations (field reseach) report number C037/96 1996 RIVO DLO.

Willoughby S. (1999) Manual of Salmonid Farming: Fishing News Books, Blackwell Science, Oxford.

Worm, B. (2016) Averting a global fisheries disaster. Proceedings of the National Academy of Sciences. 113. 201604008. 10.1073/pnas.1604008113.

World Bank (2013) FISH TO 2030 Prospects for Fisheries and Aquaculture at 4–5 www.fao.org/docrep/019/i3640e/i3640e.pdf.

World Ocean Review 2 report (2013) The Future of Fish – The Fisheries of the Future 2013.

Yussefi, M. (2010) The world of organic agriculture: statistics and emerging trends 2008. Earthscan.

# Part III

# The impact of livestock production on people

# 9 Livestock production and human rights or how the chicken came home to roost

*Raj Patel*

I was invited to use the title "Livestock production and human rights", and while the topic is vital, it could sound like a bit of a downer. Indeed I could show you a graph of bad things and explain how they're getting worse (Figure 9.1).

And then, because this graph conjures the ghost of Thomas Robert Malthus, we'd discuss population. We might observe that if we were to live with the resources that the average Bangladeshi consumes annually, we would be well within the planet's "carrying capacity" but if we lived like Americans such as myself, we would require 4.1 planets (Global Footprint Network, 2012). But I'm not here to praise Malthus – I'm here to bury him. Malthus declares that it is something about humans that has brought us to this moment. What I'll be showing you is that it's not about humans so much as the systems under which most humans live that has brought us to this edge.

To do that, I'm going to change the title of this contribution to "How the chicken came home to roost".

If we are thinking about the sort of stratigraphic markers, the things that will be part of the fossil record, then radioactivity from atomic tests and plastic in the ocean will count. But so will chicken bones (Carrington, 2016). *Gallus gallus domesticus* is the world's most popular bird. The chicken we eat today is very different from those consumed a century ago. Today's birds are the result of intensive post-war efforts, drawing on genetic material sourced freely from around the world, which humans decided to recombine to produce the most

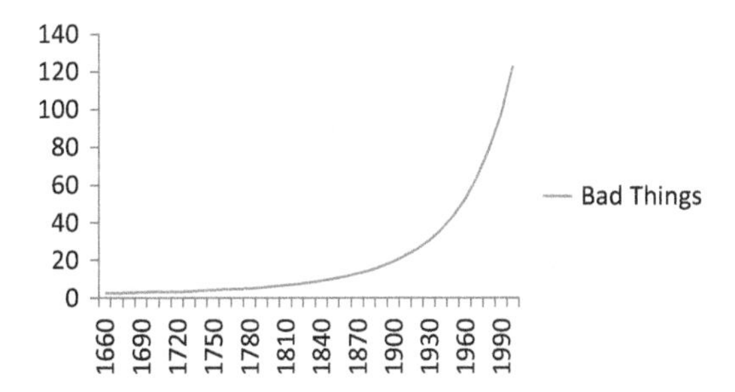


*Figure 9.1* "Bad things" representation.

profitable fowl. That bird can barely walk; reaches maturity in weeks; has an oversized breast; and is reared and slaughtered in geologically significant quantities, already more than 60 billion birds a year (Evans, 2014). It's clear that one marker of our time is going to be the trillions of chicken bones that are laid down that are a sign that humans were here.

If you look at chicken consumption in the US it is sort of flat for a while, and then after the war it shoots up. And that's precisely because of the explosion in industrial farming that happens after the Second World War. And which has been exported planet wide. Now if we look at world poultry consumption we see that (it's another one of those graphs that's going up and getting worse) we are now at about 14 kilograms of chicken per person per year planet wide (Figures 9.2 and 9.3).

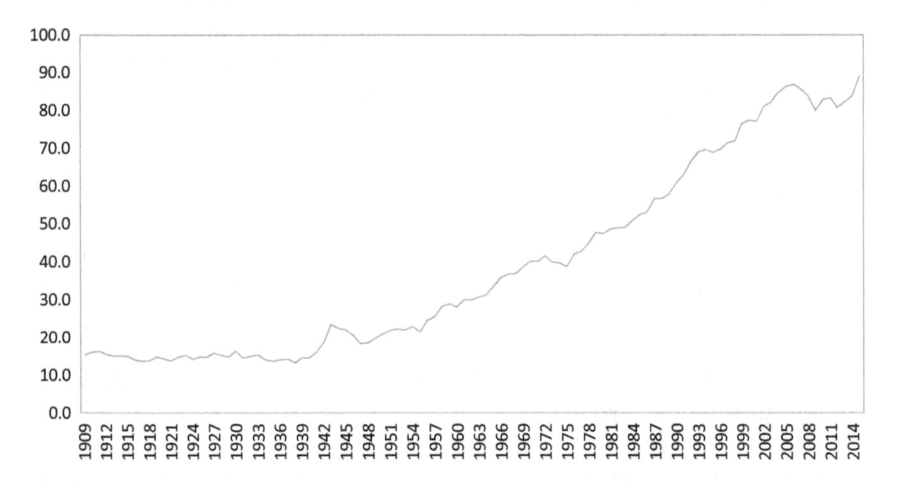


*Figure 9.2* US Total Chicken Availability 1909–2015 (ready-to-cook carcass in lbs/person).

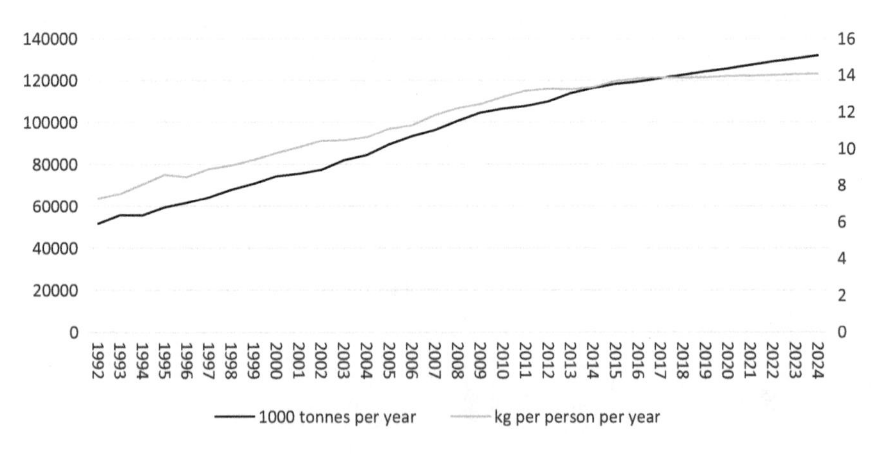


*Figure 9.3* World Poultry Consumption;
Source: OECD (2017), Meat consumption (indicator).

I want to explain what's behind this, using the idea of cheap things. Earlier on we were talking about cheap food, and I want to suggest that there are in fact seven cheap things that are worth keeping an eye out for. And perhaps the most subtle of them is the idea of cheap nature, that the world is out there for humans to do with as we will, that there are species like the red junglefowl, the ur-chicken, which we can take and mine for its genetic material, and turn into something that is so vast that its breasts are so large that often it can't even pick itself up. Yet for those species for which we can't find a profitable use, the fate is worse – 50% of the ancestral diversity of chicken species is now absent from commercial lines (Muir et al 2008). Cheap nature is about an attitude between humans and the rest of the natural world that allows us to take it for free or for very little.

But the chicken doesn't become a McNugget by itself; it requires labour, it requires work. I will talk a lot about the US because in many ways industrial farming was invented there. Industrial farming through the 1800s happens in the US. The stock yards of Chicago are the epicentre, the invention of mass production of meat. Already the most popular meat in the US, chicken is projected to be the planet's most popular flesh for human consumption by 2020 (Bunge, 2015). For that to happen requires a great deal of labour. Poultry workers are paid very little; two cents for every dollar spent on a fast-food chicken goes to workers in the US, where some chicken operators use prison labour, paid 25 cents per hour.

So, in the US, we have been pioneering the use of cheap work by workers whose bodies can be subjected to occupational illnesses at rates five times higher than those at regular jobs, seven times higher rates for carpal tunnel and ten times higher rates for repetitive strain injury (Oxfam America, 2015).

The cheap work isn't only part of the chicken business – it is actually part of the food business. In the US, the Bureau of Labour Statistics compiles data about how much people earn. Arrange a list of occupations by ascending median hourly wage, and you'd start at $9.27 for gambling dealers to, 33 jobs later, $10.58 for farmworkers and labourers. In between, 17 of the lowest-paid jobs are food system jobs, from waitstaff to pot washers to cooks to workers in nurseries and greenhouses. The food system is premised on the idea of people working very cheaply and it's not just in the US that this happens. A report was released by the International Labour Organisation recently, saying that there are 40 million people working in the conditions of modern-day slavery, 5.4 victims of modern slavery for every thousand people in the world in 2016 (International Labour Organization, 2017).

It's important that 71% of modern-day forced workers are women. Look at the sorts of work that people in forced labour and slavery do. Agriculture accounts for around 11%, but the largest part is domestic work – care work. How is that linked to our story about chicken? Well, when the bodies of workers are spat out of the assembly line who takes care of them? In the US we have incredibly expensive healthcare. Cheap care is a prerequisite for capitalism to run. In a patriarchal society it often falls to women to perform this work. When the United Nations Development Programme ran the numbers in 1995, they found the world's total product was $33 trillion dollars but $16 trillion was uncounted unpaid work, of which $11 trillion was women's unpaid work (UNDP, United Nations Development Programme, 1995).

Cheap care isn't the only thing that is required to make a chicken nugget. We need cheap food. If you are being paid a minimum wage, then all of a sudden a dollar burger seems attractive. Financially it makes sense that cheap food is part of a world where cheap work and cheap care have been naturalised. The Overseas Development Institute observes a general trend that processed food is increasing in price far less than fresh fruit and vegetables. Cheap food is necessary in order to have workers who are paid very little (Wiggins et al 2015). When we bemoan cheap food, we can't simply pine for food to be more expensive because doing so will hurt the poor disproportionately. We need, as Philip Lymbery explains in his chapter, to breach the silos, we need systemic thinking.

Cheap food is the corollary not just of cheap work and cheap care and cheap nature but cheap money. In order for KFC to exist in the US, the industry needs low interest loans and concessions. Franchises are considered small businesses and they are given loans up to $2 million by the Small Business Administration (Guides website). This is in addition, for instance, to the cheap feed that the birds get, a subsidy equivalent to $1.25 billion a year (Starmer et al. 2007). Cheap money through subsidy and tax breaks is what makes it possible for nearly half the exploitation of oil in the US to become profitable at $50/barrel (Erickson et al 2017). Cheap energy is of course part of the story. In order for chickens not to freeze and die, you need access to propane, cheap fossil fuel to heat the barns.

And perhaps the most pernicious part of this story of cheap things is that there are certain classes of people who are considered disposable. Look at who works in the US poultry industry: disproportionately women and the people of colour on the production line. In the US there is a long history of white supremacy, though the UK has to deal with its rather distasteful racial history too.

So, what do you do about that? Well, here's an awkward bit of data which KPMG, in their 2012 report, called *Expect the Unexpected – Building Business Value in a Changing World* (Cooperative, KPMG International, 2012). Using some fairly conservative assumptions, they said let's look at the output of every industry; the earnings before interest; taxes, depreciation and amortisation; the big revenue number. Then let's look at the environmental externalities of this industry and compare them as a proportion of revenue. They looked at the airline and automobile industry, and they found that even in the fossil fuel industry there are 760 billion dollars of revenue, and the externalities were 23%. So, even if the fossil fuel industry fuel did right by the planet, there would still be money left over. But not in the food industry. This is the exception. The food industry had a turnover of $89 billion dollars and its externalities were 224% of revenue. Not of profits, but revenue. I was sent to this report by one of the vice presidents of sustainability at Nestlé who corroborated these ratios. So, when the VP of Sustainability at Nestlé says this number is right, what is he saying? He is saying that under current assumptions of labour and environmental externalities there is *no way* for industrial food to be sustainable. The only way we make money is by making you pay for it or by exploiting the planet or by exploiting indigenous people. Using the seven cheap things. I worry that if we ask our business partners to step up and do the right thing, they will do the things that are most profitable for them and which might be a win for society. But the biggest wins aren't when they make money; it's when they lose and we win. So what do we do about that? I think we have to change and understand what the system is.

We shouldn't call our era of mass extinction the Anthropocene – because this is not about just people being people and humans in the way that fish will be fish or boys will be boys. Not all human civilisations have destroyed the planet. It's only some humans, some civilisations. So, let's call it not the Anthropocene but the Capitalocene. But this era doesn't have to end in chaos and the end of all life. There are groups – not individual consumers – but groups pushing back. Regarding cheap lives we can talk about black lives matter, but actually native lives matter is an important movement throughout the planet. Groups pushing a vision for indigenous people have been part of this. If we think about energy, a lot of sustainable energy advocacy is looking for a system change. If we are interested in the end of cheap nature – and this is a way of answering Carl Safina's question (Chapter 11) about what it looks like to respect nature in a way that isn't exploitative – ask the coastal Salish People and the Haida people. They have a salmon festival each year where the first salmon is harvested and then celebrated for ten days. During that festival no one fishes, so in the ten days that you are celebrating your treaty with the salmon people, the rest of the salmon are spawning. There is a relationship between you and the salmon which is about recognising your dependence on this animal. But then letting it flourish, and building into your society a relationship that is not about exploitation but about treaty and exchange. That is a very different way of thinking about nature at the moment, but we have these examples in the US. The US has 500 nations in it with many of these kinds of examples. When it comes to work we can imagine not being subject to bosses but working cooperatively. When it comes to care, unionising care work is again a feature that is happening around the world. And when it comes to cheap money, maybe reparations is an idea we need to talk about in order for the poorest people in the world to be able to get over this population hump. To conclude, then, we have no shortage of inspiration to make change away from a world based on seven cheap things towards one that recognises the web of life that sustains us all. But in order to move to that world, we need to name capitalism and work hard to build what happens after it.

This chapter is based, with permission of the University of California Press, on the introduction of *A History of the World in Seven Cheap Things*, by Raj Patel and Jason W Moore, 2017.

## References

Bunge, J. 2015. www.wsj.com/articles/how-to-satisfy-the-worlds-surging-appetite-for-meat-1449238059.

Carrington 2016 Working Group. 2016. Working Group on the 'Anthropocene'. http://quaternary.stratigraphy.org/workinggroups/anthropocene/, accessed 21 September, 2016. Here we refer to the Anthropocene in its strict geological sense. This Geological Anthropocene is, however, distinct from the problematic Popular Anthropocene. See the essays collected in Moore (2016).

Cooperative, KPMG International. 2012. Expect the Unexpected: Building Business Value in a Changing World. KPMG.

Erickson, Peter, Adrian Down, Michael Lazarus, and Doug Koplow. 2017. "Effect of subsidies to fossil fuel companies on United States crude oil production". *Nature Energy*. doi: 10.1038/s41560-017-0009-8.

Evans, 2014. www.thepoultrysite.com/articles/3230/global-poultry-trends-2014-poultry-set-to-become-no1-meat-in-asia/.

Global Footprint Network, 2012. See www.footprintnetwork.org/.

Guides.wsj.com/small-business/franchising/how-to-finance-a-franchise-purchase/.

International Labour Organization. 2017. Global Estimates of Modern Slavery: Forced Labour and Forced Marriage. Geneva: International Labour Organization.

Muir, William M et al. 2008. "Genome-wide assessment of worldwide chicken SNP genetic diversity indicates significant absence of rare alleles in commercial breeds". Proceedings of the National Academy of Sciences 105 (45): pp. 17312–17317. doi: 10.1073/pnas.0806569105.

Oxfam America. 2015. Lives on the Line: The Human Cost of Cheap Chicken. Washington DC: Oxfam America.

Starmer, Elanor, and Timothy A Wise. 2007. Feeding at the Trough: Industrial Livestock Firms Saved $35 billion From Low Feed Prices. Tufts University, Medford MA: Global Development and Environment Institute.

UNDP, United Nations Development Programme. 1995. Human Development Report 1995: Gender and Human Development. New York: Oxford University Press.

Wiggins, Steve, and Sharada Keats. 2015. The Rising Cost of a Healthy Diet: Changing Relative Prices of Foods in High-Income and Emerging Economies. London: Overseas Development Institute. doi: 10.1787/fa290fd0-en.

# 10 The role of livestock in developing countries

*Jimmy Smith*

## Key messages

- Many of the Sustainable Development Goals (SDGs) cannot be achieved without the livestock sector, which plays a major role in achieving these goals.
- All countries strive for safe, sustainable livestock production that makes a positive contribution to the SDGs worldwide.
- But the entry points and transition pathways for livestock development differ markedly between the rich and the poor, between the developed and developing worlds.
- We must not allow the concerns and interests of the rich to jeopardise those of the poor.

## Introduction

In 2015, the global community signed up to the Sustainable Development Goals (SDGs) – 17 ambitious goals that describe the world that all aspire to achieve by 2030. Unlike their predecessors, the Millennium Development Goals, which focussed on addressing the most intractable challenges in the developing world, the SDGs apply globally. While actions to achieve them will differ from country to country and region to region, they are goals to which every nation now subscribes. The role of the livestock sector in developing countries can thus now be set against this broad backdrop. It is imperative to consider the global drivers and trends for the sector, the opportunities for its contributions to sustainable development and the implications of livestock being excluded from the development narrative and actions.

## The role of livestock in developing countries must be set in the context of global trends anticipated in the coming decades

Demand for animal-source foods is predicted to rise significantly over the coming decades (Robinson & Pozzi 2011; OECD & FAO 2016). Almost all the increased demand will take place in developing countries, where it is being driven by increased population, small rises in income and considerable urbanisation – not by excessive consumption. (Developing-country per capita consumption of meat and milk is likely to remain one-third to one-half that of developed countries even by 2050.) The Organisation for Economic Co-operation and Development (OECD) estimates that in 2016 the average annual meat consumption in the European Union was 69 kg/capita compared to just 8 kg/capita (OECD 2017) in sub-Saharan Africa. The rising demand for animal-source foods in developing countries opens new opportunities for national economic growth, given the major role the livestock sector plays in the national economies of these countries – contributing 40% of agricultural GDP and growing (De Haan 2016; Pradère 2014). In these developing regions where demand for animal-source foods is growing, these foods are being produced largely by smallholder farmers (Hemme & Otte 2010; Lapar *et al.* 2012) whose livestock are playing a vital role in the household economies of some three-quarters of a billion of the world's poorest people. Today and into the coming decades, over half of the world's meat, milk and cereals is produced by small-scale men and women farmers in developing countries (Herrero *et al.* 2013).

It's noteworthy that most of the developing world's staple cereals could not be produced without manure (Liu *et al.* 2010), draught power (Starkey 2010) and income from animals. The transition of this smallholder sector to one able to meet the rising demand for livestock foods in ways positive for the environment, for livelihoods and for health is an opportunity not to be missed.

## There is risk that this opportunity will be missed if misperceptions about the livestock sector continue to be promoted without nuancing and fact-checking the messages

Like other sectors, the livestock sector faces many challenges, with those related to its harmful environmental impacts and the danger of overconsuming animal-source foods often being headlined. Undoubtedly, these are issues to be recognised and tackled. Research has an important role to play in identifying ways of doing so that don't at the same time jeopardise the potential benefits the sector provides. A brief consideration of just three dimensions will illustrate this point:

- Biodiversity and habitats

  The destruction of both plant and animal biodiversity and habitats is often highlighted as a negative impact of the livestock sector. However, there are many opportunities for well-managed livestock grazing at moderate levels to enhance the biodiversity of wildlife and plants alike (Gerber *et al.* 2013). This is especially true across the world's vast rangelands, which make up some one-third of the ice-free land of the planet. Domesticated food-producing animals are often blamed for consuming grain that could be eaten by people directly, but a recent study shows that 57% of land used for feed production is not suitable for crop production and 86% of the dry matter eaten by animals cannot be consumed by humans (Mottet *et al.* 2017). Several recent studies have also considered the links between diets and land use and find that diets including some animal-source foods use less land than vegan alternatives (Peters, Picardy, Darrouzet-Nardi, Wilkins, Griffin & Fick 2016; Van Kernebeek, Oosting, Van Ittersum, Bikker & De Boer 2016). The concept of a "circular bio-economy" recognises the essential roles of animal agriculture in using what is often "waste" (crop byproducts and the like) to produce food (Fresco 2017).

- Greenhouse gas emissions

  Greenhouse gas emissions from livestock make up about 14% of all human-induced emissions (less than transport, energy and industry) and 30% of total agricultural emissions. Like the transport sector, there are considerable opportunities to make the livestock sector more efficient rather than to discard it and there are crucial roles in this work for research-for-development organisations. These include options to make short-term productivity gains, which can reduce emissions per unit of product (Gerber *et al.* 2013) as well as longer-term ambitions such as "low-carbon cows" (Bell, Ekhard & Pryce 2012). Many countries have recognised that agriculture and, within it, livestock have important roles to play in contributing to nationally appropriate mitigation actions (NAMAs), but for this to be realised, accurate information on current emissions is needed. At present, estimates for developing-country greenhouse gas emissions are based on developed-country data; more accurate figures for developing countries may be considerably lower, as exemplified in preliminary data for emissions from livestock manure applied to soil (Pelster *et al.* 2016).

  The livestock sector plays a role not only in generating greenhouse gas emissions but also, especially in developing countries, in strengthening household and community resilience against climate and other shocks. Animals are often the one asset that people living in the driest, most marginal environments can rely on. Securing these assets, such as through novel insurance products, can reduce people's vulnerability to recurring drought (https://ibli.ilri.org/).

- Nutritional importance of animal-source foods, especially for those with little choice of diet

  Whilst for those of us living in abundance, overconsumption of animal-source foods may be a factor contributing to an increase in non-communicable diseases, for about 750 million people who are undernourished, animal-source foods play vital roles in improving nutrition. It is noteworthy that 30% of the world's protein comes from animal-source foods, which also provide vitamin B12, micronutrients and more bioavailable macronutrients than plant-based foods. There is an increasing body of evidence that animal-source foods are essential in the first 1,000 days of life (Grace *et al.* 2018), with milk and eggs especially important for children's growth and for preventing stunting (Hoddinott, Headey & Dereje 2015; Iannotti *et al.* 2017) (currently affecting 155 million children globally) and meat especially important for improving the cognitive ability of children (Neumann, Murphy, Gewa, Grillenberger & Bwibo 2007). There is no moral equivalence between those who make poor food choices and those who have no choice of foods.

## Excluding livestock from the development narrative and actions jeopardises achieving the Sustainable Development Goals

Directly or indirectly the livestock sector impacts all 17 of the SDGs. The sector's contributions to eight of the goals are most direct (Ballantyne 2017), and for these, excluding livestock from development dialogues and actions jeopardises their achievement. Four goals exemplify this point:

- *SDG1, no poverty:* Given that three-quarters of a billion people currently rely on livestock for their livelihoods, excluding animal agriculture would mean reducing their chance to escape poverty.
- *SDG2, food security:* Food security is intimately dependent on livestock, which provide both nourishing milk, meat and eggs and the inputs needed for food crop production.
- *SDG3, good health and well-being:* If livestock are excluded from the sustainable development agenda, we will not eliminate stunting in over 150 million children, malnutrition in more than 800 million people and micronutrient deficiencies in some two million.
- *SDG13, climate action:* Livestock present the biggest opportunities to mitigate greenhouse gas emissions from agriculture.

## Conclusion

The roles of the livestock sector in development need to be considered in a pluralistic, holistic way that takes account of the diversity of the sector and the vastly different actions that are needed to ensure that its contributions to sustainable development are realised. It is essential that such potential is not marginalised by our failing to nuance livestock narratives. We must ensure that the many central roles livestock play in development are not lost by those who, from a position of abundance, highlight the sector's challenges. Ultimately, a safe, sustainable livestock sector that makes positive contributions to development worldwide is possible. The actions to ensure such a transition look very different depending on today's contrasting starting points, which range from large-scale, efficient, industrial production and overconsumption in the developed world to small-scale, integrated, often inefficient and multi-purpose livestock and mixed crop-and-livestock production in much of the developing world.

## References

Ballantyne, P.G. (2017). *Achieving multiple benefits through livestock-based solutions:* Summary report of the 7th Multi-Stakeholder Partnership meeting of the Global Agenda for Sustainable Livestock, Addis Ababa, 8–12 May 2017. Nairobi, Kenya: International Livestock Research Institute. http://hdl.handle.net/10568/88091.

Bell, M., Eckard, R. and Pryce, J. (2012). Breeding dairy cows to reduce greenhouse gas emissions. In: Javed, K. (ed). *Livestock Production.* London: InTech. doi: 10.5772/50395. Available from: https://mts.intechopen.com/books/livestock-production/breeding-dairy-cows-to-reduce-greenhouse-gas-emissions.

De Haan, C. (2016). *Prospects for livestock-based livelihoods in Africa's drylands.* World Bank Studies. Washington, DC: World Bank. doi: 10. 1596/978-1-4648-0836-4.

Robinson, T.P. and Pozzi, F. (2011). *Mapping supply and demand for animal-source foods to 2030.* FAO Animal Production and Health Working Paper 2. Rome: Food and Agriculture Organization of the United Nations. Available from: www.fao.org/docrep/014/i2425e/i2425e00.htm.

Fresco, L. (2017). *Can livestock production meet the growing demand for meat in developing countries?* The Borlaug Blog. Available from: www.worldfoodprize.org/index.cfm/88533/18099/can_livestock_production_meet_the_growing_demand_for_meat_in_developing_countries.

Gerber, P.J., Steinfeld, H., Henderson, B., Mottet, A., Opio, C., Dijkman, J., Falcucci, A. and Tempio, G. (2013). *Tackling climate change through livestock: A global assessment of emissions and mitigation opportunities.* Rome: Food and Agriculture Organization of the United Nations.

Grace, D., Dominguez-Salas, P., Alonso, S., Lannerstad, M., Muunda, E., Ngwili, N., Omar, A., Khan, M. and Otobo E. (2018). *The influence of livestock-derived foods on the nutrition of mothers and infants during the first 1,000 days of a child's life.* ILRI Research Report 44. Nairobi, Kenya: International Livestock Research Institute.

Hemme, T. and Otte, J., (2010). *Status and prospects for smallholder milk production. A global perspective.* Rome: Food and Agriculture Organization of the United Nations.

Herrero, M., Havlík, P., Valin, H., Notenbaert, A., Rufino, M.C., Thornton, P.K., Blümmel, M., Weiss, F., Grace, D. and Obersteiner, M. (2013). Biomass use, production, feed efficiencies, and greenhouse gas emissions from global livestock systems. *Proceeding of the National Academy of Sciences.* 110, pp. 20888–20893. doi: 10.1073/pnas.1308149110.

Hoddinott, J., Headey D., and Dereje, M. (2015). Cows, missing milk markets, and nutrition in rural Ethiopia. *Journal of Development Studies.* 51(8), pp. 958–975. doi: 10.1080/00220388.2015.1018903.

Iannotti, L.L., Lutter, C.K., Stewart, C.P., Gallegos Riofrío, C.A., Malo, C., Reinhart, G., Palacios, A., Karp, C., Chapnick, M., Cox, K., and Waters, W.F. (2017). Eggs in early complementary feeding and child growth: A randomized controlled trial. *Pediatrics.* 140(1): art. no. e20163459. doi: 10.1542/peds.2016-3459.

Lapar, M.L., Toan, N.N., Staal, S., Minot, N., Tisdell, C., Que, N.N. and Tuan, N.D.A. (2012). *Smallholder competitiveness: Insights from pig production systems in Vietnam.* Presentation at the 28th triennial conference of the International

Association of Agricultural Economists, Foz do Iguaçu, Brazil, 18–24 August 2012. Nairobi, Kenya: International Livestock Research Institute. http://hdl.handle.net/10568/21780.

Liu, J., You, J., Amini, M., Obersteiner, M., Herrero, M., Zehnder, A.J.B. and Yang, H. (2010). A high-resolution assessment on global nitrogen flows in cropland. *Proceedings of the National Academy of Science.* 107(17), pp. 8035–8040. doi: 10.1073/pnas.0913658107.

Mottet, A., de Haan, C., Falcucci, A., Tempio, G., Opio, C., and Gerber, P. (2017). Livestock: On our plates or eating at our table? A new analysis of the feed/ food debate. *Global Food Security.* 14(Jan), pp. 1–8. doi: 10.1016/j.gfs.2017.01.001.

Neumann, C.G., Murphy, S.P., Gewa C., Grillenberger, M., and Bwibo, N.O. (2007). Meat supplementation improves growth, cognitive, and behavioral outcomes in Kenyan children. *Journal of Nutrition,* 137(4), pp. 1119–1123.

OECD. 2017. Meat consumption. Paris: OECD. Available from: https://data.oecd.org/agroutput/meat-consumption.htm.

OECD and FAO (2016). *OECD-FAO Agricultural Outlook 2016–2025.* Paris: OECD. Available from: doi:10.1787/agr_outlook-2016-en.

Pelster, D.E., Gisore, B., Koske, J.K., Goopy, J., Korir, D., Rufino, M.C. and Butterbach-Bahl, K. (2016). Methane and nitrous oxide emissions from cattle excreta on an East African grassland. *Journal of Environmental Quality.* 45(5), pp. 1531–1539. doi: 10.2134/jeq2016.02.0050.

Peters, C.J., Picardy, J., Darrouzet-Nardi, A.F., Wilkins, J.L., Griffin, T.S. and Fick, G.W. (2016). Carrying capacity of U.S. agricultural land: Ten diet scenarios. *Elementa Science of the Anthropocene.* 4, 116. doi: 10.12952/journal.elementa.000116.

Pradère, J.P. (2014). Links between livestock production, the environment and sustainable development. *Scientific and Technical Review of the Office International des Epizooties.* 33(3).

Starkey, P. (2010). *Livestock for traction: world trends, key issues and policy implications.* AGA Working Paper. Rome: Food and Agriculture Organization of the United Nations.

Van Kernebeek, H.R.J., Oosting, S.J., Van Ittersum, M.K., Bikker, P., and De Boer, I.J.M. (2016). Saving land to feed a growing population: consequences for consumption of crop and livestock products. *International Journal of Life Cycle Assessment.* 21(5), pp. 677–687. doi: 10.1007/s11367-015-0923-6.

# Part IV

# The significance of sentience in wild and farmed animals

# 11 Beyond words

## What animals think and feel

*Carl Safina*

I am going to answer one question, a question that many of us have asked ourselves, and I'm sure you will recognise it: "Does my dog really love me – or does she just want a treat?"

By looking at their sweet faces we can *easily* tell that they *obviously* really love us, can't we? Or maybe it's not so easy.

Let's consider, first, whether we have the right question. When we ask whether "they love us", that's not really a question about them. "Do you love me?" is a question about me. I needed a very different question. The question I came up with is "Who are you?"

We are aware of the capacities of the human mind, such as consciousness, intelligence, empathy, love – but are these capacities only of the human mind? What are the other big brains that share our planet doing with what's inside their heads? Elephants and apes, dolphins, birds, fish – are they completely devoid of mental experience? Are they just machines? It doesn't seem so.

But some scientists have for a long time told us that *nothing* is going on even in the largest brains on the planet. They've said that the question itself isn't scientific because you cannot find a way in.

That might have been true 60 years ago, but it is not true now. There are at least three very good ways in. You can look at brains, you can consider evolution and you can watch what they do.

Jellyfish were the first creatures with nerve cells. Jellyfish gave rise to chordates. Chordates gave rise to vertebrates, which came out of the water and had conferences on animals and sustainability and things that we need to know about. But it's still true that a nerve cell is a nerve cell whether it's in a jellyfish or chimpanzee or your dog or a crayfish. It is still a nerve cell.

So what does that tell us about the emotional and cognitive lives of something like a crayfish? It tells us very little. But it turns out that if you give crayfish in tanks little electrical shocks every time they try to come out to look for food, they will develop what looks like an anxiety disorder. They will stop foraging and appear to shut down and seem depressed, as though they are frightened. If you add to the water a drug used to treat anxiety disorder in humans, chlordiazepoxide ("Librium"), they relax and come out again (Fossat et al. 2014).

Now, how do we consider the possibility for anxiety in crayfish? Mostly we boil them by the thousands.

Octopuses are molluscs but they recognise human faces and they use tools as well as do most apes. How do we honour the ape-like intelligence of this mollusc?

Mostly we boil them by the thousands.

On reefs there are fish called groupers. Sometimes groupers chase smaller fish into crevices in coral or rocks, and when they do that the grouper sometimes goes to where it knows a moray eel is sleeping. It has a way of signalling to the moray "follow me". The moray understands that signal, follows the grouper, which appears to direct the eel to where the fish is hiding and the moray goes in. Sometimes the moray catches the fish, and sometimes the fish bolts, and the grouper gets it (Vail et al. 2013). This is an ancient partnership, which has probably been going on for millions of years, although we have only learned about it quite recently.

How do we honour this ancient partnership? Mostly fried.

Now a pattern is developing and this pattern is saying more about us, unfortunately, than it says about them.

Teaching is when you take time away from what you are doing to show a companion or a youngster how to do something. Humans do a lot of teaching. Chimpanzees, interestingly, do not teach. But otters teach (Bender et al. 2009), and killer whales do a lot of teaching (Matkin et al. 2011), and killer whales share all of the food they catch. Some of their teaching, about certain evolved hunting techniques, takes several years to complete.

If you look at the human brain compared to a mouse brain, you see that the human brain is an elaboration on the basic mammalian brain. This is how evolution works; it doesn't just come up with something totally new. It takes the parts that are in stock or on the shelf and then it fabricates some new twist. So if you compare a human brain with a chimpanzee brain, you can see that the human brain is basically a very large chimpanzee brain. It's a good thing that we have the biggest because we are the most insecure of all of the apes. Yet, if we just consider size, the neocortex (the thinking part) in a bottlenose dolphin brain is larger than it is in a human brain (Bearzi et al. 2008). Does this tell us anything about how dolphins think or feel? It might. It might not.

We can see brains but we can't see minds. Yet we can see the workings of minds in the logic of behaviours. Elephants are often seen to choose a cool spot in the shade of the trees to let their infants go to sleep while the big ones do not sleep; they rest and they doze but they stand vigilant facing out and touching one another. We can see exactly what choice they have made. How is it that our minds make sense of this exactly as their minds do? Why would we choose exactly how they have chosen? Because for a very long time under the arc of the same sun, listening to the same roars and whoops of the same enemies, we became who we are and they became who they are. We have been neighbours for a very long time.

Researchers have done experiments where they played recorded human voices in different languages to see how elephants react. If elephants hear the voices of tourists, they ignore them; tourists don't bother elephants. Yet when they hear Maasai voices, they bunch up and run away because the Maasai carry spears and they get into confrontations with elephants at waterholes, and elephants get hurt. Elephants know that not only are there different kinds of animals and some are human; they know there are different kinds of humans and some are okay and some are dangerous. They know us better than we know them. They have been watching us for a long time, and more carefully than we've watched them. They know us better than we know them.

We look different on the outside but our imperatives are the same: try to stay alive, find food and protect our babies. Under the skin we are almost identical: the same skeleton, the same organs. Whether we are outfitted for hiking in the hills of Africa or for diving under the ocean, mammals are basically the same. In the flippers of killer whales there are the same finger bones that we have. We have almost identical nervous systems. The same hormones and chemicals that create mood and motivation in human brains and minds are found in their nervous systems and their brains as well. And so you see things in elephants and other social mammals like helping where help is needed; like curiosity, especially in the young; and you see the deep bonds of family affection. Affection is affection. Dancing is dancing. It is what it looks like.

And then we ask a really weird question: "Are they even conscious?" Well, when you are given general anaesthesia, what do you become? Unconscious. It means you are disconnected from the sensory experience of your sense organs. To ask this question of animals who have eyes that can see, noses that can smell and ears that can hear, and who play with each other – is very weird.

Why are we still asking it? Because it reinforces our favourite story. Our favourite story is that we are special, we are the best, we are the only ones who matter. But there are many other creatures out there who are what I call "who" animals. "Who" animals know *who* they are. They know who their family and friends are. They know who their enemies are – or might be. They make alliances and learn to cope with rivals. They may challenge the existing order. They often aspire to higher rank in the group; their status may affect the prospects of their offspring. They are defined by their personal relationships. Does all this sound familiar? We are just one species among other "who" animals.

People say empathy is the thing that makes us uniquely human. Yet empathy is just the ability of your mind to match the emotional state of others. That's empathy. All animals that live in groups need empathy because when it's time to hurry up, you'd better hurry up. If all of your companions suddenly startle and fly away, it does not pay for you to stand there and say, "Gee I wonder why everybody just left". That's not going to work.

I think there are three realms of empathy on a sliding scale. One is the basic empathy I just mentioned, basic mood matching. The next has a slight distance that I call sympathy, such as "I am sorry to hear your grandmother has passed away". I don't feel the same grief, but I understand that grief. If you are moved to do something about it, that is what I call compassion; acting on sympathy is compassion. Human empathy is not "the" thing that makes us human; in fact it is very far from being perfect. We take empathetic animals and round them up; we kill them, and we eat them. If you don't think that is a good example then all I can say is that we don't always treat each other too well either. Slavery, trafficking, war and violence – we are demonstrably unable to gain full control over human impulses to harm other humans.

People who know one thing about animal behaviour usually know one very awkward technical term: anthropomorphism. It means attributing human thoughts or emotions to non-human beings. And we are told that we are not allowed to do so. But it's not scientific to say they're hungry when they are looking for food and to say they're tired when their tongues are hanging out but to say, when they are playing with their young and having a good time, "I wonder if they experience anything at all, I wonder if they are even conscious?" That is not scientific. We deny them their joy to reinforce our favourite story, that we are the special ones that matter.

I was talking to a reporter who said "OK, you are telling me all this stuff, but how do you really know that other animals can really think and really feel?" I was searching my brain for the best published example and then I realised the answer was right there on the carpet. When my puppy wants her tummy rubbed, she does not go over to the dining table to roll around on her back – she comes to me and rolls over on her back. She does that because she has just thought, "I would like my belly rubbed. I'm going to go to him because we are family. I trust him completely and I know that when I roll over on my back he knows what I am asking and he knows how to get the job done and make it feel good". She has thought, she has felt; it's not too much more complicated than that, really.

We see wild animals, and we say, "Oh look they're elephants, they're killer whales, they're wolves". We put a label on them and think we've seen them. That's not how they see each other. Take the killer whales in the picture (Figure 11.1), who are all members of the L-pod, off Washington State in British Columbia. That tall-finned male was 36 years old when I took that picture, and that female next to him is his cousin, who was 44. They've known each other for decades. They have travelled many of thousands of miles together. When they are miles apart in the ocean they hear each other and know who is who. They know who they are with and where they are, who their enemies are, they know who their friends and families are. They have lives.

*Figure 11.1* Orcas of the L-pod. Copyright Carl Safina.

Humans are not only capable of feeling grief, we generate an incredible amount of it. Take the case of elephants. We want to carve their teeth. Why can't we wait for them to die? Elephants once occupied Africa from the Mediterranean to the Cape of Good Hope. Now they live in little disconnected shards and we are grinding those shards smaller and smaller all the time. In the picture is an elephant named Philo in Kenya in the Samburu reserve. The next picture shows Philo 4 days later, a bloody carcass with tusks cut off (Figure 11.2).

Of course in the US, where I come from, we take much "better" care of our wildlife. In our most famous Park we paid the Park rangers to completely exterminate the wolves, some of the last wolves that were living south of the Canadian border. The last wolf in Yellowstone National Park was killed by a ranger in 1926. There had been hundreds of thousands of wolves across the US; then there were none.

Without predators, the elk population surged. Their growing population began to devastate the vegetation in the Park. Between 1930 and 1970, the Park shipped and killed thousands of elk. By the mid-1990s, they brought wolves back to try to bring the deer and the elk under control again. Nowadays, people coming to Yellowstone just to see the wolves spend 30 million dollars every year. But in 2012 the US Congress took wolves off the Endangered Species List. They went from being completely protected to being completely unprotected once they put a paw outside that Park.

Just a couple of months after that happened, two members of the best-known wolf family in Yellowstone were fatally shot. Immediately the younger

*Figure 11.2* Left: Filo alive. copyright Ike Leonard. Right: Filo 4 days later. Copyright Carl Safina.

wolves started to fight with each other. One of the most precocious females was kicked out of her family by her sisters. The father was wandering around, trying to find his mate and his brother, who had been killed; he lost his territory and his family – his hunting support – going into winter. I thought the female would be okay, and he was doomed. Instead what happened was that she was shot, starving at a chicken coop; he stayed alive for years. Sometimes the consequences of killing animals are greater for the ones who *survive* because when their families are broken up the trajectory of their lives can change. We have forgotten that they have lives; they are not just numbers (Figure 11.3).

We cause them so much pain that I often wonder why it is that they don't hurt us more than we hurt them. Killer whales have never hurt a human being in the wild. I've seen a killer whale who has just been eating part of a grey whale that he and his companions had killed, swimming alongside a small boat. The people had nothing to fear. Killer whales eat seals; they don't eat people. Why not? Why can we trust them around our toddlers? Why is it that two different scientists in two different countries have similar stories about being lost in the fog and being taken to their houses on the shore by killer whales? These are mysteries.

In the Bahamas scientist Denise Hersing has been studying spotted dolphins for 30 years and knows about 100 of them by name. They know her, recognise her boat and they show up when she arrives. One day she got there and they wouldn't come near the boat. She could not understand what was

*Figure 11.3* Wolf 755, who lost his mate and brother, who had been killed. Copyright Alan Oliver

wrong with the dolphins. Then someone came out from below deck to announce that a person on board had just died during a nap in his bunk. Did the dolphins somehow know that one of the human hearts had stopped? Why would that spook them? The scientists left to take the man back to shore and do the sad business that they had to do. When they came back a few days later all the dolphins came to the boat as usual.

What is the explanation for all these strange events? I don't know. But I know what it means. It means there is a lot more going on in the other minds that share our planet than we ever suspect.

In South Africa there was a baby dolphin of nursing age, named Dolly. At the aquarium where she was living, one day one of the keepers was smoking a cigarette while watching her through the window. She went over to her mother and nursed for a moment or two, came back to the window and released all the milk so that it encircled her head like smoke. When human beings use one thing to represent something else, we call that: art.

The things that make us human are not the things we tell ourselves are the things that make us human. Our major components are not really so different from so many other beings. Many do things that we do; we do things that they do.

What I think makes us human is that we are the most *extreme* of all the animals. We are the most creative and the most destructive and the most compassionate and the cruellest animal that has ever lived. That's us. We are all of those things together. But we are not the only ones who love each other, we are not the only ones who care for our mates or for our children.

Albatrosses sometimes fly for a month to go 8,000 miles to bring back one meal for their chick. They live on the most remote islands in the oceans; we don't even know that they are there, but boy do they know we are there. And now what do they feed their children after flying for a month, 8,000 miles? I have found dead fledglings full of cigarette lighters, toothbrushes and screw tops to plastic jars.

This is not the relationship we are supposed to have with the rest of the living world. But it is the relationship we do have because we – who have named ourselves for the minds we are so proud of – don't use those minds to think about the consequences of what we do.

When we expect new human life in the world, we paint *animals* on the walls of the nurseries. We don't paint cell phones or work cubicles. We paint animals. We don't even realise why we paint animals. But I think it's because we have this unconscious blessing ready for our new babies: "Welcome to this beautiful living world. We are not alone, we have company". But every one of those animals on the wall of the baby's room and every animal in every painting of Noah's Ark – deemed worthy of salvation by the Creator – is in mortal danger now. And their flood is us.

We started with a question. The question was "Do they love us?" And we are going to end by turning that question around and asking, rather simply, "Are we capable of loving them enough to simply have them continue to exist?"

Many of the animal encounters described in this paper are covered in depth in Carl Safina's book "Beyond Words – what animals think and feel". First published 2015 by Henry Holt and Co, New York. Paperback Publisher: Picador, 2016. Published in the UK by Souvenir Press Ltd. Also available in: Kindle, eBook, Audio. ISBN: 9781250094599.

## References

Bearzi, M., and C. B. Stanford. 2008. *Beautiful Minds: The Parallel Lives of Great Apes and Dolphins.* Harvard University Press, Cambridge, Massachusetts, 351 pp. ISBN-13: 978-0-674-02781-7.

Bender, C., et al. 2009. "Evidence of Teaching in Atlantic Spotted Dolphins by Mother Dolphins Foraging in the Presence of their Calves" *Animal Cognition* 12(1): 43–53.

Fossat, P., et al., 2014. "Anxiety-like Behaviour in Crayfish is controlled by Serotonin" Science 344: 1293–1297.

Matkin, C., et al 2011. "Killer Whales in Alaskan Waters" *Whalewatcher* 40(1):24–29

Vail, A. L., Manica. A., & Bshary, R., 2013. "Referential gestures in fish collaborative hunting", Nature Communications, Vol. 4, 2013, p. 1765.

# 12 Creatures of the factory farm

*Joyce D'Silva*

## What are we *doing*?

Isolating, caging and confining, mutilating, force-feeding or semi-starving, breeding for unhealthy levels of productivity – the long list of the horrors we inflict upon factory-farmed animals is both shocking and totally unjustifiable. It also reflects a 16th-century Cartesian outlook that animals do not have the capacity to suffer – a view now known to be absolutely incorrect and unscientific.

## Extinction and growth

Elsewhere in this book, experts have shown how industrial farming harms the environment, removes habitat for many wild species, poisons creatures both directly and indirectly, for example, through pesticides or pollution of their habitat. Taken all together, this kind of farming leads to many species finding it hard to survive and sometimes to extinction.

So while the numbers of species and individual wild creatures is shrinking and threatened, the exact opposite is happening to the animals we farm. Their number is growing exponentially. Currently we slaughter over 70 billion farmed animals a year – excluding fish. Fish slaughter estimates put the numbers of wild fish caught each year from 1999 to 2007 at between 0.97 and 2.7 trillion, and the numbers of farmed fish at 37–120 billion (in 2010) (http://fishcount.org.uk/fish-count-estimates). These numbers are also on the increase.

It is impossible to "put a face" on numbers like these. Yet each one of those billions IS an individual, with a life that matters to them and a capacity for pain and suffering. But let's not dwell only on their capacity to suffer – by corollary they have a capacity to enjoy a decent life, to exist in a state of fundamental well-being. This state, which doesn't preclude incidental or periodic adverse states, is what we expect for ourselves and desire for our loved ones and others. You or I may break an arm, which hurts, but we get treatment and return to our homes and may get help from family and friends. We may suffer a relationship break-up or a bereavement – again we usually have support and help from others. The suffering may be intense but we adjust and become able to enjoy other pleasures in the future.

## The other kind of family planning

The animals whom we keep in factory farms enjoy no such "goods". Their lives are planned well before birth by the industry and are based on economic analyses, such as "if we lose 5% of the chickens in the shed before their slaughter age of 6 weeks, can we still make a profit?" Such calculations neglect to include the suffering of the chickens, who are bred for fast and meaty growth and who may become severely lame, be in constant pain and be unable to reach the food and water in their sheds or who may drop dead from acute cardiac problems, known as "Sudden Death Syndrome" (Figure 12.1).

Breeding for high productivity is the unseen cruelty in factory farming. Those chickens are bred to reach slaughter-weight in half the time it took chickens 50 years ago. Their bones cannot support the great weight of muscle (meat) that they are carrying around (de Jong, I.C. et al., 2011). But it's not just chickens who suffer. The modern black-and-white dairy cow with Holstein genes has been designed for maximum milk yield, sometimes producing ten times the amount of milk her calf would have suckled from her.

Although she is likely to suffer from lameness and often painful mastitis too, she will usually survive to produce two or three calves before her health conditions become chronic and too expensive to treat and she will be culled (Whay, H.R. et al., 2003). She will still be slaughtered at about a quarter

*Figure 12.1* Broilers are crowded together and bred to grow fast (Compassion in World Farming/Martin Usborne).

of her natural life span (Forbes, D. et al., 2013). I remember a hard-nosed veterinary surgeon telling me that the saddest thing he saw at cattle markets was the contrast between the young heifers, with their bright eyes and shiny coats, and the cull dairy cows, who are bony, often limping or with diseased udders and a depressed look about them, although they are only three years older than their bright-eyed sisters.

Talking of cows raises the much-ignored issue of maternal nurturing – or rather, the lack of it – in industrial farming. Those dairy cows are allowed to suckle their newborn calves for only up to about 24 hours so that the calf receives health-promoting colostrum. The calf is then removed so that the cow can be milked to capacity for commercial profit. Some calves never get enough colostrum, which makes them more vulnerable to infections. The male calves may be reared for veal or for dairy beef and the females reared as "herd replacements" for their worn-out mothers. If you watch the calves being taken away from their mothers, you will see the deep distress it causes to both. The mothers try to follow their calves and bellow loudly – and often for a long time (Figure 12.2).

It's not just dairy cows who have their offspring removed. Piglets are also weaned off their mothers at four weeks old, whereas in natural conditions they would suckle for three months or so. But this premature weaning ensures that the mother breeding pigs can come into oestrus quickly and be impregnated again for the next profitable batch of piglets. Early weaning of piglets is associated with higher levels of antibiotic use (Jorgensen, D., 2013).

Of course neither the chicks of laying hens nor broilers ever get to see their mothers. They are hatched in huge hatcheries and never know how their mothers would have taught them, by specific clucks and behavioural example, which foods to eat, which to avoid and generally how to behave in a chicken-like way.

*Figure 12.2* Excessively large udders on a dairy farm.
Credit: Compassion in World Farming

## Gender issues in the factory farm

In a feminist age it seems apt to look at gender issues in industrial faming. The female chicks of laying hens get reared to be egg-producers. The males are slaughtered straight away as they are the wrong breed for meat. Their gassed or suffocated bodies are minced up for feeding to other animals such as pet or zoo-captive snakes. A whole swathe of low-paid workers act as "sexers", inspecting the tiny chicks on the conveyor belt and deciding their fate based on their gender.

In the dairy industry, it's usually the male calves who get reared for veal, as the females are destined to replace their worn-out mothers. Although the European Union (EU; including the UK) have banned the keeping of veal calves in crates so narrow that they cannot turn round, calves can still be reared in individual pens for the first eight weeks of their lives – they just have to have room to turn round (Council Directive 97/2/EC). The narrow crates are still in use elsewhere. For naturally sociable animals this isolation must be hugely frustrating. Young calves – like lambs – love to play in groups. On happier free range farms you can see groups of calves racing about, apparently chasing each other or, maybe, just for the joy of it.

## Off the land and into confinement

Obviously animals in factory farms have to be fed, as they are not outside, where they would be able to graze, browse or root for their own food. It's amazing how many issues arise because of that simple act of separating animals from the land and providing them with food in confinement. We know from other chapters in this book how the growing of cereals and soy for animal feed can have devastating impacts on wildlife habitats and biodiversity. In addition the process of feeding these foods to the factory-farmed animals, whilst boosting fast growth and milk yield, raises welfare issues. High-protein cereal-and-soy diets are associated with acidity and can have knock-on effects on the feet of cattle, predisposing them to lameness.

To produce foie gras, regarded as a delicacy, the ducks or geese have to be subjected to very invasive procedures. For the last weeks of their lives they are force-fed fatty corn by having it rammed down their throats with a mechanical augur (or hand-and-stick). As intended, their livers cannot cope and swell up, creating the aptly named foie gras or "fatty liver".

## Breeding for profit and pain

Although broiler chickens are fed generously to speed up their growth rates, the breeding birds who are their ancestors, suffer from the opposite scenario – semi-starvation, with small amounts of feed provided just once every 24 hours. Why? These breeder birds grow fast – they've been genetically selected to do so, so that their offspring will grow fast too. But if the breeder birds ate like their descendants, they would be unlikely to reach their own puberty at 18 weeks of age, as the majority would go lame or suffer cardiac problems. Many could not live long enough to breed. So their feed is severely restricted, which must cause them considerable anguish. Compassion in World Farming took this issue to the High Court in London in 2003. The judge agreed that these breeder birds were in a state of "chronic hunger", whilst not ruling in our favour for a change to the law (Neutral Citation, 2004).

Farm animals may also be subjected to a variety of painful mutilations – to make them fit the unnatural rearing systems in which they are housed. Pigs often have their tails docked, usually without anaesthesia or pain relief. In a barren concrete pig pen, the bored and frustrated pigs may turn to biting the wiggly tails of their pen-mates. Infection could set in and expensive antibiotics might be needed. To prevent this, farmers frequently routinely dock the tails of their piglets, although EU law requires them to seek other solutions first, such as enriching the environment (European Commission, 2008). Piglets reared outdoors in free range systems have no interest in each other's tails – there are far more interesting things to do! (Figure 12.3)

There is a long list of other mutilations which may be inflicted on pigs and poultry, most associated with assaulting animals' bodies in order to make it profitable to keep them in crowded and confined conditions (Figure 12.4).

*Figure 12.3* The maternal bond is strong.

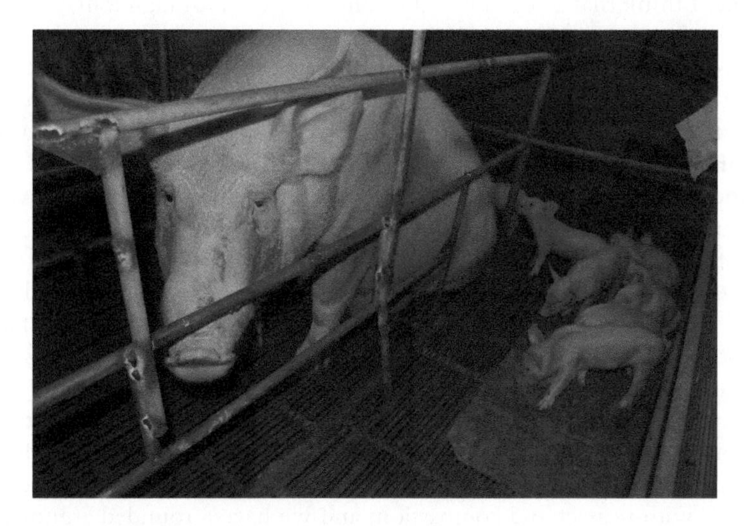

*Figure 12.4* Sow and piglets in a farrowing crate. The maternal bond is frustrated.

So, we now have a situation where billions of factory-farmed animals suffer:

> Psychological and social deprivation, being taken from their mothers and/or kept isolated from their peers and/or being crowded together in such numbers that normal group relationships become impossible.
> Psychological frustration and deprivation from being kept in uncomfortable and barren housing conditions.
> Physical pain from breed and productivity-associated conditions such as lameness and mastitis as well as pain from mutilations and from inappropriate flooring, such as concrete, metal slats or wire.
> Feed-related problems from the wrong sort of food, lack of it or force-feeding.

Emotional deprivation from being unable to experience feelings of joy and well-being because of all of these.

## Conclusion

We now look back on slavery and child labour as unjustifiable and abhorrent practices (although they still exist in some parts of the world). I believe that at some future point we shall look back at industrial animal farming in the same way – how COULD we have treated our fellow sentient beings like that?

We may reach a point where we feel any kind of farming and slaughter is unacceptable – some believe that a vegan world may come about. Looking around me, I think that a vegan world is some way off, although it may come about in the future.

What is urgently needed now is legal reform to ban the worst practices associated with factory farming. Alongside that, we need companies to speed up the path which many have begun, to require higher welfare standards from their suppliers and to encourage more plant-based eating. We need religious and social leaders to join the call for reform and institutions to divest from companies which perpetuate cruelty.

Most of all we need people to make ethical dietary choices, to speak out and demand change from politicians so that governments promote healthy eating. We need consumers to urge retailers to stock only higher welfare animal products and to promote the growing number of alternatives to animal-based products.

This book has made a strong and urgent case for abandoning industrial farming on environmental grounds. The case for the health benefits of reducing meat consumption has been made. Add to these the ethical case for treating animals with respect and compassion, and we have a rounded argument to adopt, promote and live by. I urge you, the reader, to do so!

## References

Council Directive 97/2/EC of 20 January 1997 amending Directive 91/629/EEC laying down minimum standards for the protection of calves. https://publications.europa.eu/en/publication-detail/-/publication/fc4da6bc-4781-44ec-97e5-54d3693d3f70/language-en

de Jong, I.C., Moya, T.P., Gunnink, H., van den Heuvel, J., Hindle, V.A., Mul, M.F. and Van Reenen, C.G., 2011. Simplifying the Welfare Quality assessment protocol for broilers= Vereenvoudiging van het Welfare Quality protocol voor het meten van welzijn bij vleeskuikens (No. 533). Wageningen UR Livestock Research.

European Commission, 2008. Council Directive 2008/120/EC of 18 December 2008 laying down minimum standards for the protection of pigs. *Official Journal of the European Union L* 47, 5–13.

Fish Count: http://fishcount.org.uk/fish-count-estimates, accessed 26/2/18.

Forbes, D., Gayton, S. and McKeogh, B., 1999. *Improving the longevity of cows in the UK dairy herd.* Cirencester, Milk Development Council.

Jorgensen, D., 2013. www.ft.dk/samling/20131/almdel/flf/spm/495/svar/1156714/1401964.pdf

Neutral Citation Number: [2004] EWCA Civ 1009, Case No: C1/2003/2699, in the Supreme Court of Judicature, Court of Appeal (Civil Division), on Appeal from the Queen's Bench Division, Administrative Court, The Hon Mr Justice Newman, CO/1779/2003.

Whay, H.R., Main, D.C.J, Green, L.E. and Webster A.J.F., 2003. Assessment of the welfare of dairy cattle using animal-based measurements. *Veterinary Record* 153, 197–202.

# 13 The scientific basis for action on animal welfare and other aspects of sustainability

*Donald M. Broom*

## Sustainability

A key question about any production system, including those where it is human food that is being produced, is whether or not it is sustainable. The meaning of sustainability is now much wider than it was in early writings on the subject because the ethics of the production method are now included, and a system can be unsustainable because of negative impacts on human welfare, animal welfare or the environment. A definition of sustainability is a system or procedure is sustainable if it is acceptable now and if its expected future effects are acceptable, in particular in relation to resource availability, consequences of functioning and morality of action (Broom 2014, modified slightly after Broom 2001, 2010).

The concept of the quality of goods that people buy has also been changing. Whilst quality still includes immediately observable aspects and the consequences of consumption, for many people it now includes the ethics of the production method. Consumers now require transparency in commercial and government activities and take account of the ethics of food production when they evaluate product quality (Broom 2010, 2017).

What makes a food production system unsustainable and results in product quality being judged as poor? The impacts of each of these aspects of production can now be measured in an objective scientific way.

a Adverse effects on human welfare, including human health

Food products are not just evaluated on taste and price. If they cause people to become sick, the quality is considered poor. Some foods are regarded as being better for the health of the consumers because of the nutrients present in them. A major effect on animal production in recent years of attempts to provide a healthy diet has been the dramatic increase in the production of farmed fish, in part because they contain polyunsaturated fats (Wall et al. 2010). Another human and animal health impact is that in all aspects of farming, the use of antibiotics and other antimicrobials will have to decrease, in most countries via new legislation. The development of antimicrobial resistance (AMR) means that many of these drugs are no longer effective, partly because of misuse of antibiotics in human medicine but partly because of widespread rather than just therapeutic use in livestock farming (Ungemach et al. 2006). It is estimated that 65% of antimicrobial usage by weight is for farm animals (EU data from ECDC/EFSA/EMEA/SCENIHR 2009). Because the animals are a mean of 2.4 times the weight of a human, calculations using the European Union (EU) data indicate that the number of treatments with antimicrobials is 44% for farm animals. For the antimicrobials assessed by World Health Organisation (WHO) as the most important for humans, the figures for animals are 36% by weight and 19% by treatments. Every human patient and every farmer in the world should respond to this dangerous situation. There is still use of antimicrobials in farming that substantially increases the risk that resistance will develop. As with human use, the risk is much greater in some countries than in others, especially in those where the antimicrobial can be obtained by users without prescription by a medical doctor or veterinarian and where prescribers are easily persuaded to prescribe when they should not. The use of antimicrobials as growth promoters, a practice that is becoming illegal in more and more countries, is a particularly damaging practice.

b Poor welfare of animals

Many consumers will not buy animal products if there is close confinement of animals, individual rearing of social animals such as pigs and cattle, and other systems for housing and managing animals that do not meet the needs of the animals. An increasing result of this is the number of people who decide to become vegetarian or vegan. Other people just decide not to buy particular animal products. Hence some widely used animal housing systems are unsustainable (Broom 2017). Animal welfare is a key aspect of sustainability and product quality.

c Unacceptable genetic modification

The use of genetically modified plants is not accepted by some consumers and few people accept the use of genetically modified or cloned animals. All cloning of farm animals is associated with poor welfare of animals, and this is the reason why it is not permitted in the EU (Broom 2014, in press). The public's antipathy to genetic modification and cloning is partly dislike of modifying what is natural. Another aspect is that modified organisms may have allergenic proteins and many of the public do not believe that proper checks on such possibilities are in place (Lassen et al. 2002). Animals which are genetically modified may have welfare problems so there should be checks, using a wide range of welfare indicators, before they are used in any way (Broom 2008, 2014).

d Harmful environmental effects

Agricultural methods that result in low biodiversity are a consequence of widespread herbicide and pesticide use and are perceived to be the norm by many farmers and some of the general public. However, such an impact is far from inevitable and there are agricultural systems that lead to biodiversity on farmland being much increased. Livestock production can also result in pollution locally and on a worldwide scale, for example, via greenhouse gas production. Greenhouse gas production can be reduced by modified feeding and land management systems. Maintaining resources, such as soil with good structure, and retaining water that might be lost from the soil are important objectives, as are minimising usage of carbon-based energy and imported fertilisers. Soil is often damaged by tillage and the emission of greenhouse gases (Pagliai et al. 2004). There has been over-exploitation of all open water fish and of whales and widespread extinction of species is occurring very rapidly now. In some cases this occurs because of a specific use, for example, feathers, ivory or rhino horn, but whole habitats are disappearing because of human activity. A livestock farming component that has led to a dramatic environmental effect is the widespread death of vultures in India caused by the use of the veterinary drug diclofenac (Green et al. 2004). The population declined to only 3% of its former level but is starting to recover following legislation. In temperate and tropical countries a dramatic example is that in the last 20 years we have seen the greatest decline in farmland birds, butterflies, bees and wild plants ever recorded. This was reported by Donald et al. (2000) and van Dyck et al. (2009), and it was concluded by Newton (2004) that the decline is principally because of the use of herbicides that destroy plants upon which wildlife depends, both in and close to agricultural fields. However, pesticides, especially neonicotinoids, also reduce the numbers of birds (Hallmann et al. 2014). These examples raise the question of what we want in our environment. Do we need vultures in India or farmland birds and butterflies, and hence biodiversity on farmland and nationally in our country?

e Inefficient usage of world food resources

At present, there is often very inefficient usage of food and energy resources. Much human food used in homes, sold in restaurants and sold in shops is wasted. Some food for farmed animals is wasted. Almost all of this waste could be prevented. In addition, much food that humans could eat is fed to animals that are then eaten by people. This is a much less efficient process than for the humans to eat the food directly. What can be done in animal production to exploit existing resources better (Herrero et al. 2010)? The most important animals for food production are those that eat food that humans cannot eat. Hence herbivores eating forage plants, not cereals, are much more important than pigs or poultry which compete with humans for food (Broom et al. 2013). Similarly, herbivorous fish are more important than those fish that eat other fish. Land used for agriculture is sometimes degraded because of poor management, for example, repeated tillage and use of the same crops, so it is not exploited efficiently. Too much energy from fossil fuels is used in cultivation and transport of feed and products, as well as in production of fertilisers and other materials and equipment.

f Not “Fair trade” – producers in poor countries do not receive a fair reward

Consumers in many countries have now discovered that producers of food in poor countries are often not properly rewarded for their work. Most profits from the sale of some basic products bought by many people have been found to go to large companies. This is considered morally wrong by most consumers and, as a consequence of publicity about unfairness to poor producers, products such as coffee, cocoa and fruit are among those that are independently checked and have a Fair Trade label (Nicholls and Opal 2005). Hence the producers receive a larger part of the money paid by shoppers.

g Not preserving rural communities

Small-scale rural farmers are often out-competed by large-scale production, with the result that local communities disappear. The general public often find this unacceptable so schemes are introduced by governments to safeguard such communities. Consumers may also buy locally produced products, regarding this as a part of product quality. In the EU, subsidies to preserve rural communities have prevented rural people migrating to towns and hence large cities becoming ever larger (Gray 2000; Broom 2010).

## Welfare and health

The term welfare is used for all animals but not for plants or inanimate objects. The welfare of an individual is its state as regards its attempts to cope with its environment (Broom 1986) so welfare varies over a range from very good to very poor and can be measured scientifically. This state includes all coping systems, including behavioural and physiological body regulation, the immune system and many other systems that are largely controlled by the brain. Those systems that cope with pathology contribute to health which is an important part of welfare. Positive and negative feelings are adaptive mechanisms that are central aspects of welfare. Animals with the level of awareness and cognitive ability necessary to have feelings are said to be sentient (Broom 2014, 2016).

The concept of health and the concept of welfare are exactly the same for humans and non-humans. This point is emphasised in the “one health” and “one welfare” discussions (García Pinillos et al. 2016). Many measurements are the same for humans and non-humans. Welfare measures, such as those of behaviour, physiology, immune system function, clinical condition and body damage, are described by Blokhuis et al. (2010), Broom (2008a, 2014) and Broom and Fraser (2015). Emergency adrenal responses are the same in a frightened person or sheep, or a fish taken out of water. All can result in immunosuppression. One response to pain in people is the grimace response in which the eyes are partly closed, the mouth is moved and the cheek muscles are clenched. The same response to pain is shown by sheep, goats, horses, mice and rats so grimace scales have been developed for assessing pain in these species.

As a result of much study of animal welfare science, we now have information about the needs of the main farm animal species; thus, consideration of these needs is the first step in evaluating systems for keeping and managing animals. For animal welfare scientists and legislators, this approach has largely superseded the less precise five freedoms approach. The most important animal welfare problems all concern farmed animals: broiler chicken welfare, dairy cow welfare, laying hen welfare, pig welfare and the welfare of farmed fish. There are many scientific publications on the welfare of all of these animals in the various possible production systems (Fraser 2008; Broom and Fraser 2015; Broom 2017). The first step when there is a requirement for laws or codes of practice concerning the use of animals is to have a report produced by unbiased scientists. After an objective scientific report has been produced, its conclusions and recommendations can be discussed with stakeholders (i.e. those with a financial interest in the area) and other interested parties. Legislators can then decide on laws, and both non-governmental organisations (NGOs) and companies that produce or sell food can decide on codes of practice. In relation to animal welfare in the EU, the European Food Safety Authority (EFSA) Panel on Animal Health and Welfare produces the scientific reports and makes them available on the internet. The welfare of hundreds of millions of animals has improved as a result of EU policies and legislation.

## Food production systems for the future

What is the future for the production of food and other goods in the world? Consumers in more and more countries have concerns about healthy food, biodiversity and animal welfare. There is an ever-increasing number of people with the view that we must provide for the needs of the animals that we keep and that we must use world resources more efficiently. In order to do this we should consume more plants and fewer animals. If grain is produced, it is more efficient for people to consume the grain directly than for it to be fed to animals, with much loss of energy in the process, and then to consume the animals. Where meat is consumed, it should come mainly from animals eating food that humans cannot eat. Hence we should concentrate on producing herbivorous mammals, birds, fish, etc. As a consequence, ruminants, that get their nutrients from leaves, are much more important than pigs or poultry that compete with humans for cereals and soya.

A question which arises as a consequence of such arguments is "should we stop animal production and just produce plants?" If the basis upon which this will be decided is to do with the efficient utilisation of world food resources, the answer is no. We should reduce animal production. However, approximately 45% of land in the world is good for producing food for herbivores but not for producing plants as human food. If we stopped production of animals for human consumption, this land would produce almost no food for people. The remainder of the land would have to be farmed more intensively and there would probably be major food shortages. Of the food that is produced for human consumption, as much as 30% is wasted. In order to use that which would otherwise be wasted, some could be fed to other animals. For example, after it is treated to prevent the spread of disease, much could be fed to pigs (zu Ermgassen et al. 2016). A further factor is that most of the world is sea and there is potential for it to be better used for producing marine plants and animals for human food.

A further question to consider is "is it morally right to consume animals?" The answer to this question depends upon which moral issues are considered the most important. For some people, the main view is that it is objectionable to consume animals or animal products. I see this as principally an aesthetic question but some others do not. People for whom this is the paramount issue, will not eat animals.

A second moral argument is that "it is wrong to consume animals because we should not kill animals". However, this argument does not logically lead to vegetarianism because large numbers of animals are killed in plant production. Some are small soil animals. Others are mammals, birds and insects that we call pests. Other animals die or are prevented from living at all in order that crop production methods can be used. Per unit of human food, some animal production methods allow far more animals to survive than some plant production methods.

A third position is that we have an obligation to use animal production methods only where animal welfare is good. Where animal welfare is viewed as a part of sustainability, this position can be rewritten as all food production systems should be sustainable. Returning to the initial questions, it is clear that, if we stop or reduce animal production methods that misuse world resources, more food can be produced. We should concentrate on farming herbivorous mammals, birds, fish and perhaps insects or molluscs that can be fed grass, leaves and other plant products which humans cannot digest efficiently.

## Sustainable animal and forage plant systems

For many years we have been talking about grazing systems. The key plants have all been pasture plants. Trees and shrubs have been mainly considered as competitors for the pasture plants. Yet plant production from a mixture of herbs, shrubs and trees is much greater than from a single-layer pasture system.

Some shrubs and trees provide good food for ruminants and other animals, including herbivorous fish. Shrubs such as *Leucaena* have been used as forage for ruminants for many years. However, most animal production is still from pasture only.

Work in Colombia, Mexico and Brazil on semi-intensive, three-level, rotational, silvopastoral systems has now reached a point where revolution is starting. This is because semi-intensive silvopastoral systems with grasses, leucaena *Leucaena leucocephala* or other protein-rich shrubs and trees, often with edible leaves, produce more forage and more animal product than monoculture pasture-only systems (Murgueitio et al. 2008). In addition, the welfare of the animals is better, including less disease; biodiversity is much greater; worker satisfaction is high; soil quality, including water-holding capacity, is much increased; there is less water run-off; conserved water use is six times less than feedlot systems; there is 30% less greenhouse gas production per kg meat; there is better carbon sequestration; and the land area needed for beef production is 42% of that for feedlots (Broom et al. 2013).

Especially during dry periods when herbs and shrubs are less productive, the leaves of trees like ramón (Maya nut) *Brosimum alicastrum* can be cut and fed to livestock. Shrubs and trees that are too high for animals to reach can be cut and fed to ruminants or fish. This development can be taken up in many parts of the world now but, for the future, another step is to collect and eat the insects that feed on tree leaves. This means planting forests for farming.

## References

Blokhuis, H.J., Veissier, I., Miele, M. and Jones, B. 2010. The Welfare Quality project and beyond: safeguarding farm animal well-being. *Acta Agriculturae Scandinavica, Section A, Animal Science*, 60, 129–140.

Broom, D.M. 1986. Indicators of poor welfare. *British Veterinary Journal*, 142, 524–526.

Broom, D.M. 2001. The use of the concept animal welfare in European conventions, regulations and directives. In: *Food Chain*, 148–151. Uppsala: SLU Services.

Broom, D.M. 2008a. Welfare assessment and relevant ethical decisions: key concepts. *Annual Review of Biomedical Science*, 10, T79–T90.

Broom, D.M. 2008b. Consequences of biological engineering for resource allocation and welfare. In: *Resource Allocation Theory Applied to Farm Animal Production*, ed. W.M. Rauw, 261–275. Wallingford: CABI.

Broom, D.M. 2010. Animal welfare: an aspect of care, sustainability, and food quality required by the public. *Journal of Veterinary Medical Education*, 37, 83–88.

Broom, D.M. 2014. *Sentience and Animal Welfare*, p. 200. Wallingford: CABI.

Broom, D.M. 2016. Sentience, animal welfare and sustainable livestock production. In: *Indigenous*, eds. K.S Reddy, R.M.V. Prasad and K.A. Rao, 61–68. New Delhi: Excel India Publishers.

Broom, D.M. 2017. *Animal Welfare in the European Union*. Brussels: European Parliament Policy Department, Citizen's Rights and Constitutional Affairs, p. 75.

Broom, D.M. in press. Animal welfare and the brave new world of modifying animals. In: *Pushing the Limits of Animal Biology and Its Implications for Welfare and Ethics*, eds. T. Grandin and M. Whiting. Wallingford: CABI.

Broom, D.M. and Fraser, A.F. 2015. *Domestic Animal Behaviour and Welfare*, 5th edn., p. 472. Wallingford: CABI.

Broom, D.M., Galindo, F.A. and Murgueitio, E. 2013. Sustainable, efficient livestock production with high biodiversity and good welfare for animals. *Proceedings of the Royal Society B*, 280, 2013–2025.

Donald, P.F., Green, R.E. and Heath, M.F. 2000. Agricultural intensification and the collapse of Europe's farmland bird populations. *Proceedings of the Royal Society B*, 268, 25–29.

van Dyck, H., van Stries, A.J., Maes, D. and van Swaay, C.A.M. 2009. Declines in common, widespread butterflies in a landscape under intense human use. *Conservation Biology*, 23, 957–965.

ECDC, EFSA, EMEA and SCENIHR (European Centre for Disease Prevention and Control, European Food Safety Authority, European Medicines Agency and European Commission's Scientific Committee on Emerging and Newly Identified Health Risks). 2009. Joint Opinion on antimicrobial resistance (AMR) focused on zoonotic infections. *EFSA Journal*, 7(11), 1372, 78, doi:10.2903/j.efsa.2009.1372.

zu Ermgassen, E.K.H.J., Phalan, B., Green, R.E. and Balmford, A. 2016. Reducing the land use of EU pork production: where there's will, there's a way. *Food Policy*, 58, 35–48.

Fraser, D. 2008. *Understanding Animal Welfare: The Science in Its Cultural Context.* Chichester: Wiley Blackwell.

García Pinillos, R., Appleby, M., Manteca, X., Scott-Park, F., Smith, C. and Velarde, A. 2016. One welfare – a platform for improving human and animal welfare. *Veterinary Record*, 179, 412–413.

Gray, J. 2000. The common agricultural policy and the re-invention of the rural in the European Community. *Sociologia Ruralis*, 40, 30–52.

Green, R.E., Newton, I., Schultz, S., Cunningham, A.A., Gilbert, M., Pain, D.J. and Prakash, V. 2004. Diclofenac poisoning as a cause of vulture population declines across the Indian subcontinent. *Journal of Applied Ecology*, 41, 793–800.

Hallmann, C.A., Foppen, R.P.B., van Turnhout, C.A.M., de Kroon, H. and Jongehans, E. 2014. Declines in insectivorous birds are associated with high neonicotinoid concentrations. *Nature*, 541, 341–344.

Herrero, M., Thornton, P.K., Notenbaert, A.M., Wood, S., Msangi, S., Freeman, H.A., Bossio, D., Dixon, J., Peters, M., van de Steeg, J. and Lynam, J., 2010. Smart investments in sustainable food production: revisiting mixed crop-livestock systems. *Science*, 327, 822–825.

Lassen, J., Madsen, K.H. and Sandøe, P. 2002. Ethics and genetic engineering – lessons to be learned from GM foods. *Bioprocess and Biosystems Engineering*, 24, 263–271.

Murgueitio, E., Cuartas, C.A. and Naranjo, J.F. 2008. *Ganadería del Futuro.* Cali: Fundación CIPAV.

Newton, I. 2004. The recent declines of farmland bird populations in Britain: an appraisal of causal factors and conservation actions. *Ibis*, 146, 579–600.

Nicholls, A. and Opal, C. 2005. *Fair Trade.* Thousand Oaks, CA: Sage Publications.

Pagliai, M., Vignozzi, N. and Pellegrini, S., 2004. Soil structure and the effect of management practices. *Soil and Tillage Research*, 79, 131–143.

Ungemach, F.R., Müller-Bahrdt, D. and Abraham, G., 2006. Guidelines for prudent use of antimicrobials and their implications on antibiotic usage in veterinary medicine. *International Journal of Medical Microbiology*, 296, 33–38.

Wall, R., Ross, R.P., Fitzgerald, G.F. and Stanton, C., 2010. Fatty acids from fish: the anti-inflammatory potential of long-chain omega-3 fatty acids. *Nutrition Reviews*, 68, 280–289.

# 14 Green and pleasant farming

## Cattle, sheep and habitat

*John Webster*

The founder of Compassion in World Farming (Compassion) was Peter Roberts, a dairy farmer whose mission was to encourage systems of farming that were based on proper respect for the sentience of the animals and the life of the land. This chapter adheres strictly to these principles: it identifies major concerns, explores (briefly) the evidence relating to these concerns and suggests ways in which we, who care ***about*** farm animals and the land but depend on farmers for our food, can assist the farmers, who care ***for*** these things, to farm as efficiently, sustainably and humanely as possible. I shall examine some of the main criticisms of animal farming today; accept that all of them contain a great deal of truth; give examples of where they fall short of the whole truth; and explore ways to address these concerns for the good of the people, animals both domestic and wild, and the living environment.

### Most of those who can consume too much meat and milk

The big problem with meat and dairy products is that they are both tasty and highly nutritious. In consequence, most of those who can afford to eat these things, each too much – for our health, for the health and welfare of the food animals and for the sustained health of the planet. Moreover, the problem becomes progressively greater as more and more people acquire the means to indulge themselves. If the Chinese were to eat as much meat per capita as those in the US this would require approximately 130% of our current global harvest of grain. The conventional response of agribusiness to this inescapable truth is to advocate the principle of "sustainable intensification", which in the case of food from animals is usually interpreted as more food from pigs and poultry because they convert their feed (grains and beans) into meat and eggs much more efficiently than cattle and sheep convert their feed into meat and milk. Moreover, all this can be done in big buildings, out of sight and out of mind; thereby, in theory, leaving more of the countryside for us to enjoy. This approach has little appeal to those with concern for animal welfare. I shall show that it doesn't make much ecological sense either.

## Food that we could eat is fed to animals while the poor go hungry

It is an inescapable truth that vast quantities of land have been cleared and ploughed for the cultivation of grains and soya beans to feed to farm animals, pigs, poultry and ruminants. The obvious cause of this is inequality of wealth. It does not automatically follow that feeding farm animals is a waste of food. Traditional subsistence agriculture has always (where religions permit) involved cattle, sheep and goats to harvest food that the people cannot digest from land they do not own as well as poultry and pigs to scavenge the food that we fail to pick up or cast aside. It is instructive to examine the contribution of competitive (grains and beans) and complementary (grass and byproducts) feeds to animals in modern intensive agriculture.

Table 14.1 compares the efficiency of conversion of metabolizable energy (ME) and protein in feeds for farm animals into food energy and protein in food for humans (for further explanation see Webster 2017). Measured in terms of total feed consumption, pigs and poultry are about 25%–35% efficient, and beef are much worse (less than 10%). However, when expressed relative only to competitive feeds, beef becomes as efficient (or no less inefficient) than pig meat. The amazing dairy cow, by dint of extremely hard work, converts competitive feeds into milk protein and energy at an efficiency considerably greater than 100%. Of course, the exact figures vary according to the composition of the diet. However, it should be obvious that the greater the proportion of complementary feeds in diets for animals, the better it will be for us – and, indeed, for them.

## Intensive livestock production is incompatible with animal welfare

This is the big Compassion subject; too big to review here in any detail. My present theme is restricted to the concept of green and pleasant farming, that is, the farming of cattle, sheep and other ruminants at pasture for as long as possible: and when not outside at pasture, consuming as much conserved pasture as possible in the form of silage or hay. Table 14.2 uses the "Five Freedoms" as a template on which to construct the welfare implications of different production systems for cattle and sheep.

The modern high-yielding dairy cow undoubtedly has a hard life. Genetic selection, especially in the Holstein breed, has produced cows with peak

*Table 14.1* Feed for them and food for us (from Webster 2017)

| *Conversion efficiencies* | *Eggs* | *Pork* | *Milk* | *Beef* |
|---|---|---|---|---|
| Food energy/total ME intake | 0.33 | 0.19 | 0.42 | 0.08 |
| Food energy/competitive ME intake | 0.35 | 0.24 | 1.39 | 0.24 |
| Food protein/total protein intake | 0.32 | 0.25 | 0.28 | 0.09 |
| Food protein/competitive protein intake | 0.33 | 0.28 | 1.20 | 0.30 |

*Table 14.2* Welfare implications of different rearing systems for cattle and sheep

| | *Dairy cows* | *Beef on feedlot* | *Beef at pasture* | *Sheep* |
|---|---|---|---|---|
| Nutrition problems | +++ | +++ | +/− | + |
| Physical discomfort | ++ | + | +/− | +/− |
| Pain and injury | ++ | + | n.s. | ++ |
| Infectious disease | + | ++ | n.s | + |
| Fear and stress | n.s. | + | n.s. | + |
| Behavioural restriction | + | ++ | n.s. | n.s. |
| Exhaustion | ++ | n.s. | n.s. | n.s. |

yields around 50 litres (90 pints) per day. In these circumstances the capacity of the mammary gland to produce milk puts severe strain on the digestive and metabolic capacity of the cow upstream to provide it with nutrients. Many housing systems are profoundly uncomfortable for large, bony dairy cows and the prevalence of painful lameness in the average herd is over 20%. However, the biggest welfare problem for dairy cows (and dairy farmers) is not one covered by the Five Freedoms, namely the long-term problem of exhaustion leading to early culling. The average life expectancy of cows in the UK national dairy herd is less than three lactations. To be fair, dairy breeders have come to recognise the need to reduce the emphasis on selection for milk yield and increased the emphasis on traits designed to create a more robust cow with a longer productive life. The effects of the changes in the selection index are not yet obvious but they should come in time.

Young beef cattle being finished on feedlots (e.g. in the US) can experience severe digestive and metabolic problems associated with the feeding of high-starch diets. Nutritional upsets, mixing and crowding also constitute a high risk for respiratory disease. However the feedlot environment is only experienced by beef calves for about the last 150 days before slaughter. Most cattle on feedlot have spent most of their lives on open range with their mother cows. In this regard North American calves destined to be finished on feedlots would normally expect better welfare conditions for most of their lives than male calves destined for beef from the UK dairy herd: taken from their mothers at birth and reared for most or all of their lives in barns.

Beef production from suckler cows reared entirely or almost entirely on pasture, a common and enchanting sight in the Goldilocks zone of the south-west of England where I live, have, in my opinion a life about as good as it gets for any domestic animal, with almost total behavioural freedom and in the company of their own kind – freedoms seldom experienced by domesticated dogs and horses. Strictly on the grounds of animal welfare I attach far less guilt to eating free-range beef than cheese. Sheep too, typically experience all the freedoms of free-range living. However, ewes typically give birth well before the spring to ensure best prices and the maximum feed intake from fresh grass. Most lambs suffer from some degree of cold stress, and in severe weather there will be deaths due to hypothermia and starvation. Most infectious and parasitic diseases can be controlled but chronic painful lameness is endemic in too many flocks (Figure 14.1).

*Figure 14.1* Pasture rearing has many advantages (Photo: Philip Lymbery).

## Livestock's long shadow

The Food and Agriculture Organisation (FAO 2006) report, "Livestock's Long Shadow" (LLS), carries the message that food production from animals imposes a much higher environmental cost than arable farming for crops. Cattle and sheep are also under attack for the amount of methane, a powerful greenhouse gas, that they produce as a consequence of microbial fermentation in the rumen. Once again, there is real substance in these concerns. However, both assaults are themselves open to severe criticism because they only look at one side of the picture. The LLS approach calculates the environmental cost of different food production systems in terms of the area of land necessary to produce a given amount of food for humans. This is measured in global hectares but does not take into account the differing capacity of different classes of land to support different production systems for crops and animals. An alternative and to my mind far more elegant and useful approach to analysis of the environmental cost of different systems of food production is that known as "Emergy analysis", where emergy (Em) is a measure of the amount of the original, effectively inexhaustible source of solar energy embedded in each stage of the process (Zhao and Li 2005). This is a wickedly complex form of analysis but it has the supreme merit of identifying and distinguishing the renewable (R) resources of sun, soil and water embedded in farmland from non-renewable (NR) sources such as fuel, fertiliser, labour and imported feeds (Table 14.3).

According to the LLS approach, the environmental cost of feeding 100 people on beef is 10 times the cost of cereals and 40 times the cost of soya. The emergy approach yields a diametrically opposite conclusion. By this analysis and in this example, corn and soya are least sustainable because of their

*Table 14.3* Contrasting measures of sustainability for crops and livestock production

| | *Ecological footprint*[1] | *Emergy yield ratio*[2] | *Emergy sustainable index*[3] |
|---|---|---|---|
| Corn | 31.0 | 1.07 | 0.06 |
| Soya | 8.0 | 1.74 | 0.50 |
| Grazing cattle | 353 | 3.73 | 6.80 |

1 Global ha required to feed 100 people (Pereira and Ortega 2013).
2 Emergy yield ratio (EYR) = Em, food/Em, non-renewable resources, (F/NR).
3 Emergy sustainable index = EYR/((NR+F)/R)), (Rotolo et al. 2007).

dependence on non-renewable resources; beef from cattle grazing the pampas of Argentina are the most sustainable. The LLS approach is further limited in that it does not consider the extent to which soils are being degraded by continuous cropping for corn and soya. The attraction of the emergy approach is restricted (obviously) by the fact that there is a strict limit to the amount of beef that one can get from permanent pasture alone. I shall return to this point later.

The other big question (too big to review here in detail) is the production of greenhouse gases from ruminants. Once again, however, it is necessary to point out that methane output cannot be considered in isolation but must be set alongside the capacity of grazed lands to capture and sequester carbon. For detailed analysis of this topic see an excellent recent review by Garnett et al. (2017). In brief, well-managed pastures and especially silvopastoral systems have a considerable capacity to store carbon especially in soils that have been degraded through continuous cropping or overgrazing. In all but peaty, anaerobic soils this capacity is finite but they certainly have the capacity to store a lot of carbon while we strive to develop more sustainable systems. One last point on methane production. The current cattle population of the US is about 100 million. In 1800 the bison population of North America was estimated at 60 million. This system was eminently sustainable until the arrival of the European settlers, hell-bent on slaughter. The methane problem would not exist if the cattle population was reduced by half and reared primarily from forage. The poison is in the dose.

## Grasslands and parklands are too valuable to be supported solely by the production of food

This is not yet received wisdom, but it is my big message. Whether considered on a regional (e.g. the Uplands of the UK) or global basis, the 30%–40% of farmed land that exists in the form of pasture for grazing animals is far too valuable to be sustained solely from the sale of food. The Extinction and Livestock Conference has been primarily concerned with the conflict between agriculture and wildlife and has been most constructive when considering prospects for agroecology. The value of these lands for us humans is not only defined by their capacity to produce food but by their amenity, beauty, richness and diversity of habitat, and their ability to support a good quality of life for those responsible for the care of the land. It is also necessary to point out that wildlife have rights to life and a life worth living irrespective of their appeal (or otherwise) to our sensibilities. However, there is much more to sustainability than that personified by hedges and butterflies. Permanent pastures, parklands and the marginal uplands have a critically important role to play in planet husbandry through maintenance of carbon and nitrogen equilibria and management of water resources. Sustainable husbandry is critical to the life of these lands and all that depend on them. Too often we see the results of improper husbandry: soil erosion, carbon loss, flooding and destruction of habitat. On marginal lands and in subsistence agriculture this is driven by the need to generate income solely from the production of food.

Food is a basic need for all individuals. However, it is also a source of considerable pleasure and satisfaction for which we may be prepared to spend considerably more money than strictly necessary for our survival. In economic terms food is described as an individual good. It follows that the cost of food production should be met by the individual purchaser. In advanced and reasonably affluent societies these costs can reflect individual demand for added value products, be they organic, high welfare or of local provenance. The non-food elements of the value of the land under discussion here, whether it be arable, pastoral, silvopastoral or otherwise agroecological, are public goods, and it is appropriate, therefore, that they should be rewarded from the public purse. This principle must be central to current thinking on reform of the agricultural support policy in Europe and the UK.

I began with the statement that most of those who can consume too much meat and milk. I went on to review approaches to food production from cattle and sheep reared as far as possible on complementary rather than competitive feeds, wherever and whenever possible on green and pleasant pastures within ecosystems conducive to environmental diversity, wildlife and the sustainable management of soil and water. It is self-evident that this approach cannot sustain current levels of meat and milk production – but that has to be a good thing for our health and that of the living environment.

## References

Food and Agriculture Organisation (FAO) (2006). *Livestock's long shadow: environmental issues and options*. FAO, Rome.

Garnett, T. et al. (2017). *Grazed and confused*. Food Climate Research Network, Oxford.

Pereira, L. and Ortega, E. (2013). A modified footprint method: the case study of Brazil. *Ecological Indicators* 16, 113–127.

Rotolo, G.C., Rydberg, T. and Lieblein, G. (2007). Emergy evaluation of grazing cattle in Argentina's pampas. *Agriculture, Ecosystems and Environment* 119, 383–395.

Webster, J. (2017). Beef and Dairy: the cattle story. In *The meat crisis: developing more sustainable production and consumption*. Eds. Joyce D'Silva and John Webster, Earthscan from Routledge, Abingdon, Oxon. & New York, NY, pp. 117–138.

Zhao, S., Li, Z. and Li, W. (2005). A modified method of ecological footprint calculation and its application. *Ecological Modelling* 185, 65–75.

## Part V

# Solutions for people, planet and animals

# 15 Agroecology working in Africa

## The case of Sustainable Agriculture Tanzania (SAT)

*Janet Maro*

Sustainable Agriculture Tanzania (SAT), a local Tanzanian organisation, was founded in 2011. It grew out of the grass-roots Morogoro-based Bustani ya Tushikamane or Garden of Solidarity project that I founded with fellow agricultural students in 2009. SAT was established to address social and environmental problems caused by environmentally destructive and unsustainable farming practices, which lead to food insecurity, poverty and malnutrition caused by environmental degradation through loss of top soil, water supplies and forests. These environmental problems create economic hardships and social problems that are exacerbated during cycles of drought caused by changes in climate. Climate change is, of course, forecasted to become even worse during the next decades.

Since its founding SAT facilitators have demonstrated organic farming in Morogoro to over 3,000 farmers in 120 groups from 70 villages. SAT facilitators are a key component to all projects and programmes, teaching SAT methodologies through a network of community-based farmer groups. In direct contact with farmers, facilitators collaborate closely with village and group leaders and Agricultural Extension Officers. Through participatory learning they ensure adoption of successful technologies and their expansion to neighbouring communities. Ninety-seven per cent of farmers who have attended SAT courses report that the SAT training meant that they could improve their situation (RISD, 2017).

In 2013 SAT opened a Farmer Training Centre and demonstration farm in Vianzi to serve farmers throughout East Africa. Close to becoming self-sustaining with class fees, the Centre has hosted 2,400 farmers, extension officers and youths from all over East Africa.

With its vision to reduce social and environmental problems and to provide sustainable food for the fast-growing global population, SAT uses impact-proven strategies that are based on four holistic pillars:

- **Knowledge dissemination:** SAT works with farmers face-to-face, acknowledging their experiences and local knowledge, using participatory methods. The farmers are trained in agroecological farming practices using demonstration plots within their own villages, where they learn composting, botanical extract preparation, nutrition, soil and water conservation. The trained farmers then act as group leaders and pass on their learnings to other community members, involving adults as well as youth in the process. SAT's teaching content is relevant to farmer interests and issues. It includes entrepreneurial skills and promotes a saving and lending culture. Knowledge is also distributed via *Mkulima Mbunifu*, SAT's monthly farming magazine, published in Swahili. The magazine offers practical and easy understandable information about agroecological farming methods. Additionally, social media outlets keep SAT networks abreast of current projects and news. SAT information materials are open source and freely available for the public good.

- **Application and marketing:** Upholding the philosophy that you should practice what you preach, SAT is engaged in the whole value chain of agroecological food production including agricultural production, processing, packaging and marketing. SAT also works to build consumer awareness of healthy organic and fair food. Hence, SAT can support farmers efficiently by demonstrating the ability of agroecology to transform livelihoods in a positive way. SAT trains farmers in the tenets of organic certification for product marketing, and networks farmers with a national certification organisation, linking successful farms to organic markets. Additionally, SAT runs an organic shop in Morogoro town, giving farmers an outlet for their produce.
- **Research:** SAT collaborates with farmers and universities to create demand-driven research in the under-researched field of agroecology. In an annual Workshop for Participatory Research Design, farmers are brought together with students to discuss common problems. Students are then invited to develop relevant research topics for their degree programme, and upon selection, receive funding to support their study. Furthermore, SAT provides the national and international research community with access to agroecological farmers. It also conducts research on its demonstration farm to contribute to the scientific evidence of agroecological farming methods and their potential. All research results are published and made available to SAT farmers and networks, for example, at the Farmer Training Centre. SAT has engaged more than 100 agricultural students in research collaboration or through field practical trainings. Some of those former students are now in top-level positions, promoting sustainable and safe solutions in the field of agriculture.
- **Networking:** All gained experiences of farmers and other stakeholders are shared during national and international workshops and conferences. SAT shares grass-roots-level learnings, highlighting success stories and challenges facing the farming community in Tanzania. SAT brings together small-scale farmers and other non-governmental organisations (NGOs), the private sector, the Government and its public institutions to achieve maximum impact.

## SAT creates high impact for participants

A recent evaluation (RISD, 2017) showed promising results amongst SAT trained farmers that are worth sharing and scaling:

- By using the agroecological methods learned from the demonstration plots, farmers can increase their income by an average of up to 38%.
- Economic benefits arise through increases in production reported by 66% of farmers.
- The farm profit can be maximised through utilising on-farm inputs and hence decreasing costs for inputs purchased in shops. Sixty-one per cent of the farmers reported a reduction of costs for inputs.
- Through multi-cropping systems, farmers become less vulnerable to climate change and other vacillations in weather and markets. Seventy-six per cent of farmers reported that they now have a more balanced diet, which has a positive impact on their health.
- Due to diversification and through production of organically grown produce, farmers benefit from new market opportunities. Up to 50% of farmers reported having new market access.
- Through applied agroecological practices, farmers can revive and once again use land that had been depleted through overuse of chemicals, soil degradation and erosion. Sixty-four per cent of farmers reported that they could reuse land. In total, 91% of farmers were using erosion control measures after completing the SAT training programme, whereas 30% had used them previously.
- Farm biodiversity is protected and enhanced by maximising the use of locally available resources.
- By avoiding the use of chemicals, exposure to environmental toxins is reduced to almost zero.
- Through soil management, farmers fight erosion and reduce the water consumption. Experienced farmers reported a reduction of 59% of water used in agriculture due to increased organic matter and mulches, which help to increase the water holding capacity of soil, as well as preserving soil moisture.
- Through agroforestry practices, farmers are planting trees, reducing dependence on adjacent forestlands that are under harvest pressure.
- Through applied agroecological practices, farmers no longer depend on slash and burn practices. Research (Wostry, 2014) showed a reduction in burning by 95% of SAT trained farmers.

A key component of SAT's highly successful methodology is the Farmer-to-Farmer (F2F) approach in which farmer committees are formed and given additional training not only in agroecological farming but also in leadership, facilitation, internal control, networking and marketing. These new ambassadors then have the skills to return to their communities forming new groups to be trained. In return, they receive additional resources and training. SAT is bridging farmers through these leadership networks that reach remote and often isolated villages.

Using the F2F approach SAT has already trained 556 small-scale farmers in Morogoro Region. Additionally, SAT facilitated 400 animators from Masasi Region (South-Coast of Tanzania), who are now set to train a further 5,000 farmers in their region. SAT's External Evaluation Report (RISD, 2017) reveals that most trained farmer groups remain registered and active in their communities with members continuing to learn, share and support each other, increasing their knowledge of organic farming.

## Future plans and scaling agroecology

SAT's innovative approaches are scalable and in demand. For example, SAT has expanded its facilitation services to three further regions in Tanzania in response to requests from the Ministry of Water and Irrigation and other organisations in Tanzania.

SAT is also planning innovative, farmer-driven projects with the aim of protecting watersheds and will continue its work with the Masai community to find agroecological solutions to farmer and pastoralist conflicts.

Working with young farmers and youth groups to engage them in entrepreneurship and environmental clubs is another key activity: SAT collaborates with 33 youth groups in Morogoro and Mbeya regions.

Each year since 2014, SAT has hosted organisations wishing to learn from our approaches and experiences. For example, Agroecological Focal Points of an international organisation with project managers from eight countries from four continents and delegations from Malawi and Ethiopia. SAT plans to replicate this model to other parts of Africa and beyond by networking with potential partners and institutions.

**Box 15.1 Case study – Amina Shabani**

Amina Shabani is 45 years old and the mother of six children, four boys and two girls. She lives in *Mwanzo Mgumu* village in Morogoro. Amina had no chance of going to school and does not read or write. Her husband died five years ago, and she is responsible for the care of five children since one of her children married.

In the 1990s, as a young wife, Amina started farming maize, rice and sesame with her husband. They used to produce maize and rice for food only but sold any excess for income generation. Nowadays, since receiving her training from SAT, Amina grows crops like maize, rice, cassava and banana as well as vegetables, such as amaranth, okra, spinach and tomatoes. She also keeps poultry. Amina owns 1.5 acres of land where she grows the bananas on one acre and cassava on a half acre. She rents land to grow the maize, rice and vegetables.

"To me cassava and banana are the crops that I depend on for selling, whilst I produce the maize and rice to feed my family, though I sell any excess", says Amina. Previously, she never used to grow cassava, bananas and vegetables, but after she started taking care of her family and joined the group, she decided to produce these crops to increase her income.

> Life was very tough at the beginning but after I joined the group established by SAT in 2013, and benefitted from the trainings in agroecology, entrepreneurship, basic life skills, poultry keeping and saving and lending, I realized that it is possible to break through,

she said. She sells bananas, cassava and vegetables; the income earned is used to support her family, for home expenditure, buying books and uniforms for her children. "Compared to 5 years ago, my income has increased almost 4 times. I save the excess income in the Group Saving and Lending Scheme". With her savings Amina plans to buy land and construct a new house (Figure 15.1).

*Figure 15.1* Amina Shabani in her cassava field.
Source: SAT.

**Box 15.2 Case study – Pius Paulini**

Pius Paulini is a 52-year-old farmer in *Towelo* village, Morogoro, who is married with four children. He started farming in the 1980s: traditional farming, growing crops such as maize, beans, carrots and cabbages. Later on, he applied synthetic fertilisers and agrochemicals, reporting that he incurred high costs and depleted the soil, just like other farmers, who were forced to abandon their fields.

In 2010, he attended his first SAT training session and immediately started practicing agroecology. He made terraces, started using compost and planted beneficial trees. After conducting market research, he now sells his vegetables to Morogoro town, Dar es Salaam and Mwanza. He has become an active member of the saving and lending network as well as a farmer trainer. He has trained more than 500 farmers in Morogoro and Tanga on Sustainable Land Management.

> Through practicing agroecology, I was able to regenerate my land. On my terraced fields, which are well composted, I was able to get more than 1,000 kilos of carrots that I sold in Mwanza. I used the money to buy a motorcycle for my son Moses to start a motorcycle taxi business. My other children are still in school and I manage to pay fees for them in a private English school. In the past people saw me as a poor man with nothing but my life has changed and now they ask me to train and represent them,

said Pius. He added, "I am finalising my family house, a 4 bedrooms, self-contained with a sitting room and kitchen" (Figure 15.2).

*Figure 15.2* Pius Paulini in white cap training other farmers.
Source: SAT.

## References

RISD, An evaluation of the programme "Comprehensive Evaluation of the Bustani ya Tushikamane Program" by SAT 2009–2016 Final Report, 2017.

Wostry, A., Exploring potentials for payments for environmental services. Financial incentives and other drivers for organic small-scale farmers in the Uluguru Mountains, Morogoro. Master's Thesis, University of Vienna, 2014.

# 16 Balancing the needs of food production, farming and nature in the UK uplands

*Chris Clark and Pat Thompson*

## Introduction

The UK's mountains, moors, hills and valleys (the uplands) cover approximately one-third of the UK's land surface area. Despite their wild appearance, man has modified the uplands over millennia: trees cleared, wetlands drained, land enclosed and improved, livestock introduced and large predators systematically removed. Since the Second World War, a range of policy measures underpinned by support payments (e.g. Less Favoured Area (LFA), direct payments (so-called Pillar 1 payments made under the Common Agricultural Policy (CAP)) and agri-environment payments (Pillar 2 CAP)) have helped to sustain livestock farming in the uplands. In 2003, following a series of progressive changes, the CAP was reformed and agreement made to decouple support payments from livestock numbers. This removed the perverse incentive for upland farmers to keep more livestock than their land could sustain. In 2005, production-based subsidies were finally phased out and replaced by broader rural development payments (including so-called agri-environment schemes). The introduction of targeted agri-environment payments and the cessation of headage payments have resulted in a gradual reduction in livestock numbers (Condliffe 2009).

Throughout the uplands, hill farming, game management, forestry and energy production are predominant. Though land use and associated management (especially livestock farming and game management) has had a major impact on the upland environment, shaping the nature of the landscape through grazing and burning (e.g. creating/maintaining open habitats), large areas are of international importance. There is a network of sites protected as Special Areas of Conservation (SACs), Special Protection Areas (SPAs) (under the EC Nature Directives), National Parks and Areas of Outstanding Natural Beauty (AONBs).

These same places, home to small numbers of people living in rural communities, are also used and enjoyed throughout the year by millions of people, especially in National Parks, where access in pursuit of a wide range of leisure activities, including walking, cycling and birdwatching, underpins the rural economy. Critically, these same upland places are also the source of much of the UK's drinking water and lock up vast amounts of carbon stored as peat (e.g. blanket bog).

Because of the long history of habitat loss and degradation in the wider countryside, driven largely (though not exclusively) by changes in agriculture and forestry, the uplands are increasingly a last refuge for many species, including some formerly more widespread across the UK, such as black grouse, curlew and cuckoo (Balmer et al. 2013). These species now coexist alongside upland specialists including lesser-known rare montane invertebrate assemblages, and bryophytes including the various species of *Sphagnum* moss that drive peat formation, carbon storage and the regulation of water quality and flow.

Important populations of upland birds remain in open habitats, especially in those areas where traditional extensive livestock farming, often referred to as high nature value (HNV) farming, is still practised. Here, the current mosaic of enclosed, semi-natural pastures and meadows and associated unenclosed priority habitats such as blanket bog and limestone pavement support internationally important flora and fauna. The future of many priority species is reliant on maintaining low-intensity, environmentally friendly cattle and sheep farming or HNV, particularly of species associated with in-bye grassland (meadows and pastures) and limestone grasslands, as found in the Yorkshire Dales (England) and the Burren (Republic of Ireland) (Figure 16.1).

Maintaining environmentally friendly farming is not without economic challenge and many upland farms struggle to break even. The latest report

*Figure 16.1* Spring view of the classic pattern of walls, barns, pastures and meadows in the Yorkshire Dales (Photo: Chris Tomson).

on Agriculture in the UK (Defra et al. 2017) revealed that the average farm business income for grazing livestock farms in the LFA was £19,000. A more in-depth analysis of hill farming in England (Harvey & Scott 2017) revealed the average hill farm had a Net Farm Income of just under £10,000 with CAP payments (Basic Payment and Agri-environment payments) accounting for over 30% of revenue. Whilst there is substantial variation amongst farms in terms of their overall performance, most of these farms could not survive in their current form as commercial businesses without the CAP-related public payments. This is particularly so for those LFA farms with an emphasis on cattle production (Harvey & Scott 2017).

The UK's decision to leave the European Union in 2019, resulting in uncertainty around access to markets and farm support payments, may have a significant negative impact, particularly on those farm businesses most reliant on support payments (Cumulus Consultants, 2017). This is particularly so for those farming in some of the UK's most environmentally and culturally important landscapes.

## What do we need to do to improve the economic viability of high nature value farming?

Most within the agricultural and environmental sectors understand the role that farmers play in the uplands. Despite their importance, there are major concerns with the impact Brexit will have on the future of hill farming, not least the looming reduction in farm support payments made under the CAP. The big worry of course is the fundamental issue of farm profitability. What if there is no area support and fewer environmental payments for uplands in, say, ten years' time? How will farms and the upland landscape change? Will farm businesses fail, amalgamate, collaborate or do something else? How should government respond and what role can agricultural and environmental leaders play in helping farmers to adapt and survive?

With many hill farms struggling to make a profit, the UK's decision to leave the EC has cast further uncertainty on the future of farming in the hills. According to DEFRA, in 2014/15 half of all UK farmers failed to cover their costs of production. Even with support payments, almost 20% of farms failed to achieve a farm business income. This will almost certainly be higher on hill farms.

Some hill farmers have already begun to look at this problem. One farmer, Neil Heseltine, has been reviewing his costs and the subsequent effect on his business of significantly reducing them (see Case Study). The results speak for themselves (Table 16.1). Even though sheep stocking rates have reduced and sales halved, the return on investment (farming contribution) for sheep has increased from £500 (2012) to nearly £18,000 (2016) largely because a reduced stocking rate has resulted in a significant reduction in costs.

## Case study – Hill Top Farm, Yorkshire Dales

The Yorkshire Dales comprise some of England's most beautiful upland landscapes. Here geology and years of farming have produced an intricate patchwork of meadows and pastures, walls and barns. Whilst farming has changed, wildlife continues to thrive in those parts where farming practice is in harmony with the natural environment.

Neil Heseltine and his partner Leigh raise cattle and sheep at Hill Top Farm, Malham, in the North Yorkshire dales. The 1,100-acre family farm is entirely pasture-based overlying limestone and lies at between 800 and 1,800 feet. This is a tough place to farm, with the farm on land designated as LFA (Severely Disadvantaged Area). In 2012, the farm supported 120 Belted Galloway cattle, raised entirely on pasture, and 400 ewes with significant quantities of grain fed to the sheep over the winter to increase output.

Inspired by the ethos of the Pasture Fed Livestock Association, Neil and Leigh decided to try to reduce costs and manage the farming operation without feeding grain to the sheep over winter. This meant making the difficult decision of reducing sheep numbers to 200. The result of reducing sheep numbers and cutting the cost of supplementing winter diet with grain had a dramatic impact on the farm accounts (Table 16.1). In 2012, the 400 sheep contributed net £478 to the family's income. In 2016, half the number of sheep (200) contributed net £17,779! How was this possible? In reducing input costs (e.g. feed, labour and vet costs) and choosing to graze the sheep on available grass, the sheep operation switched from loss-making to profitable. Reducing sheep numbers also resulted in a lower impact on the environment, improved sheep health and better quality of family life.

*Table 16.1* Breakdown of costs and sales for Hill Top Farm between 2012 and 2016

| | *Sheep* | | *Cattle* | |
|---|---|---|---|---|
| Costs (£): | 2012 | 2016 | 2012 | 2016 |
| Labour | 12,000 | 750 | 1,755 | 1,260 |
| Feed | 15,178 | 3,724 | 270 | 500 |
| Haulage | 320 | 0 | 668 | 0 |
| Other costs | 11,202 | 2,600 | 3,675 | 1,318 |
| Livestock purchases | 18,900 | 7,042 | 220 | 3,040 |
| Vet costs | 1,240 | 186 | 117 | 117 |
| **Total costs** | 58,840 | 14,302 | 6,705 | 6,235 |
| **Income (sales) (£)** | 59,318 | 32,081 | 17,828 | 21,310 |
| Net contribution | 478 | 17,779 | 11,123 | 15,075 |

Farming at Nethergill Farm, also in the Yorkshire Dales, Chris Clark (lead author) has further investigated the relationship between sales output, costs and profitability. Chris now thinks about food as if it were a fuel. To improve the calorific value of naturally produced food requires more in extra energy than it will produce in calorific value. You cannot get more energy out of a system than you put in. Consequently, hill farmers struggle to correct for the disadvantages of weather, latitude or elevation by adding costs.

Chris has also determined that unlike other industries, which have a linear variable cost structure, hill farming variable costs are non-linear. Once certain limits are reached variable costs increase. For example, stocking densities are unsustainable when there is insufficient grass to support grazing livestock and additional fodder is required. The purchase of supplementary fodder increases the variable cost beyond the profitable limit. Similar and additional cost increases occur when fertilisers are used, livestock is off-wintered and where vet costs are increased as a result of high herd/flock numbers.

Hill farmers that surpass the maximum stocking rate may make more money (often for a limited time only) but will ultimately experience lower profitability (profits as a percentage of sales). For most farmers further expansions will take them to a point, where after having made profits, the business becomes unprofitable again. Paradoxically, as evident from the case study above, reducing livestock numbers (and costs) enables farmers to be more profitable.

The stock in trade for most hill farmers in the UK uplands is sheep but the truth is that lamb (as a commodity) is in long-term decline with the age demographic of lamb consumption dominated by older people. (NSA & NFU, 2014). Demographics also play a part in the entrepreneurial inertia of an aging farming population with many farmers set in their ways, thereby maintaining an unprofitable status quo.

However, change they must. Of course, every farmer's situation is different. Geology, soil type, altitude and weather all have an influence as do the needs and wants of the farming family managing the business. Hill farmers need to examine the intersection between productivity and profitability. Look at beef production, for example, where it could be three or four years before any return accrues. This presents a major management challenge for a tightly budgeted hill farm. Clearly, we need to find a more sustainable route to profitability. Depending on the business, the first step would be to look at achieving a better model, and that might be realised by a combination of:

- Reducing fixed and variable costs;
- Finding income from outside the farm;
- Local collaboration;

- Taking control of the downstream food supply chain and diversification with diversification, only when the core business is sound.

For example, a hill farmer will typically sell the product, say lamb, through a wholesaler to a meat processor, who then sells to a merchant. That merchant will set prices and sell to a retailer, who markets to the consumer. This route gives the hill farmer 15% of the final price to the consumer, while the retailer takes 35%. The merchants can take a staggering 50% of the final price (Chris Clark, personal observation).

What would happen if hill farmers worked as a collective and were able to sell direct to the retail trade? Hill farmers could form local producer groups. They could then offer products to the retailer unique to their region (e.g. all from the Yorkshire Dales National Park). With the merchant cut out, farms would take a far greater share of the profits. Working in this way might help deny supermarkets access to the group's brands, except where they are prepared to pay a premium.

The bottom line is hill farmers need to urgently plan, budget and promote in these uncertain times. They need to reassess the way they work if they are to survive and keep our uplands intact.

In contrast to the increasingly intensive livestock production systems found throughout many parts of the world, grass-fed sheep and cattle play a key role in the management of internationally important habitats and make an important contribution to the economy of rural areas in the UK uplands. Consumers can help by purchasing sheep and beef products from local producer-groups, thereby supporting the continuation of HNV farming. Such an approach is consistent with encouraging consumers to think about the provenance of what they eat and to encourage them to consume less but better meat.

## References

Balmer, D.E., Gillings, S., Caffrey, B., Swann, R.L., Downie, I.S. & Fuller, R.J. 2013. *Bird Atlas 2007–11: the breeding and wintering birds of Britain and Ireland.* BTO, Thetford.

Condliffe, I. 2009. Policy change in the uplands. In Bonn et al. (ed.). *Drivers of environmental change in uplands.* pp. 59–89, Routledge, London.

Cumulus Consultants Ltd. 2017. *The potential impacts of Brexit for farmers and farmland wildlife in the UK.* A commissioned report for the RSPB. Cumulus Consultants Ltd, Gloucestershire.

Department for Environment, Food and Rural Affairs, Department of Agriculture, Environment and Rural Affairs (Northern Ireland), Welsh Assembly Government, The Department for Rural Affairs and Heritage, The Scottish Government,

Rural & Environment Science & Analytical Services, Agriculture in the United Kingdom, 2016, Crown copyright, 2017.

Harvey, D. & Scott, C. 2017. *Farm business survey 2015/16. A summary from hill farming in England*. Rural Business Research, Newcastle University.

NSA & NFU. 2014. A Vision for British Lamb Production.

# 17 Chikolongo – a win for people and parks

*Joseph Okori and Peter Borchert*

If you were to make your way towards the village of Chikolongo in the heart of Malawi, you would be almost certain to come across elephants browsing and wallowing peacefully along the banks of the nearby Shire River. And upon entering the village you would soon learn that for a number of years people and their crops have been safe from marauding elephants and that elephant poaching incidents have been few and far between. You would also find a thriving, largely self-sufficient community with successful farming projects, safe access to resources, including fresh water, and a strong sense of civic purpose. Chikolongo is no utopia, life remains tough in this poor country where the vagaries of drought and flood can wreak havoc, but it is a far cry from the pervasive desperation and antipathy towards conservation that prevailed less than a decade ago.

What happened to create such optimism, opportunity and progress?

## Beautiful but crowded and poor

Malawi is a small, landlocked country, probably best known to most Westerners as the place from which pop star Madonna adopted two of her children. It is a place of great beauty and diversity at the southernmost reach of the East African Rift Valley, the massive geological trench that carves its way down almost the entire length of Africa. In places mountains reach some 10,000 feet, while elsewhere the land is barely above sea level.

Dominating the country is the deep, narrow, but long lake that takes its name from the country embracing its shores and waters. At its southern end Lake Malawi spills into the 250-mile-long Shire River, which feeds the shallow marshy morass of Lake Malombe before finding its way, lazily and spreading, towards the Kamuzu Barrage. Below this man-made barrier the Shire rapidly gains momentum in its tumbling, twisting rush to join forces with the mighty Zambezi in Mozambique.

Malawi's natural beauty, however, cannot mask the fact that it is one of the poorest countries in Africa and, indeed, the world. It is also very populous – upwards of 19 million people live here, more than in the neighbouring states of Zimbabwe (16.9) and Zambia (17.6), which are, respectively, three and six times bigger in land mass. This circumstance has conspired to place enormous environmental pressure on the land, with alarming consequences. The rate of deforestation is the highest in all of Africa, causing widespread soil erosion and general habitat degradation. In turn, this impacts seriously on food production and makes Malawians highly vulnerable to extreme weather events.

## A wildlife Island in a sea of people

Little wonder that when, in 2012, International Fund for Animal Welfare (IFAW)'s CEO Azzedine Downes flew from Malawi's capital to Liwonde National Park, he was struck by the sharp contrast between the lush green of the park and the surrounding community land, where barely a living tree stood standing. Liwonde was and still is, quite literally a wildlife island in a sea of people – completely surrounded by densely populated, extremely poor communities engaged primarily in subsistence agriculture. The park is one of Africa's lesser-known gems and indisputably one of Malawi's most beautiful reserves: a sprawling 212 square miles of lagoons, marshes, seasonal floodplains, open savannah, woodlands, hills, mountains and, the life blood of it all, the Shire River.

The Shire (pronounced "Shireee") forms Liwonde's western border and was largely unknown to the world outside of Africa until explored in 1859 by David Livingstone on his Zambezi expedition; the striking borassus palm trees that trim its banks are a legacy of the Arab slavers who used it as a trading route. The river is dense with rafts of hippo and some of the biggest crocodiles you could imagine. Liwonde is one of southern Africa's most important biodiversity hotspots and a major focus of Malawi's lucrative tourism industry.

## The challenge – bridging the gap

But Liwonde is under threat for it is a place where people and hundreds of elephants, rhinos, hippos, grazing mammals and over 600 species of birds battle it out daily for access to food, water and grazing space. Illegal fishing in the Shire River – a nursery for catfish and bream – had reduced fish stocks locally and upstream in Lake Malawi, affecting the fishing industry and taking its toll on the availability of a key source of protein. And as fish stocks diminished, commercial poaching of land animals grew, with hunters turning to the cruellest form of poaching – snaring animals – to find an alternate protein source. Snaring is indiscriminate, any animal can be caught not just those sought for the pot. And if the poachers don't return to clear their snares regularly, animals die slow and excruciating deaths.

"How could we convince local communities that it was in their best interest to protect elephants and other wildlife living in and around Liwonde National Park?" Downes asked, aware that here, as in many parts of Africa, communities living near national parks had not been offered much incentive other than to see conservation areas not as worthy of protection but rather as

*Figure 17.1* Trio of elephants in Liwonde National Park (Neil Greenwood © IFAW).

places of conflict between people and wildlife, where the only benefit lay in poaching and illegal fishing. Throughout Africa, and elsewhere in the world, there has been justifiable criticism of often-broken promises and attempts to impose wildlife conservation policies from far-off boardrooms, all of which have often left communities increasingly marginalised, resentful, even victimised, and feeling that they have nothing to gain from conservation.

IFAW had for some years already been active in Liwonde National Park, working towards providing a safe and protected haven for wildlife, particularly for elephants. While sensitive to community needs, IFAW's focus was then on strengthening the park's capacity to prevent and respond to threats to them and other wildlife in the park, including poaching, the bushmeat trade, human/animal conflicts and loss of habitat. It was soon apparent, however, that long-term success could not be sustained unless mutually beneficial solutions could be found for the wildlife in the parks as well as the people living around the borders (Figure 17.1).

## New beginnings

And so IFAW began a dialogue with one particular community – Chikolongo, a village of some 6,000 people on the western boundary of Liwonde and a classic example of this systemic dilemma. Antagonism towards Liwonde's elephants was rife, understandably so given that these great grey animals had long raided village crops with impunity, often causing extensive damage to meagre livelihoods and posing a very real threat to the lives of the villagers themselves. As a result poaching was also rife: elephants were being killed for their ivory and the Shire River, a central feature of the national park, was being unsustainably and illegally fished. The effects of this were being felt much further afield as well, for the river provides a nursery environment for bream and catfish that eventually swim upstream to Lake Malawi. Illegal fishermen, many using recycled mosquito nets, were taking even the smallest fish and this was seriously impacting overall stocks, ultimately posing a threat to human food security given that fish are one of Malawi's key sources of protein.

Chikolongo was Downes's final destination, the purpose of which would be to find out what would persuade the village chiefs to take a different view and to see value in helping to conserve the park's resources? The answer lay in proper partnerships with genuine participation in conservation management in which cultural heritage would be recognised and respected. Partnerships in which helping to keep elephants and other wildlife safe would also have to offer some guarantee that the villagers too would be safe, along with their crops and granaries.

The obvious priority that emerged was first and foremost to secure crops against the elephants, but the Chikolongo elders also identified the safety of people, mostly women and children as a serious concern. They faced daily danger from crocodile and hippo attacks while to-ing and fro-ing to fetch fresh water from the river. The fact that this resulted in a tragic toll of some 18 deaths every year speaks for itself. Furthermore, a fish farm was mooted as a project that could reduce the villagers' reliance on the river for food security as well as the wider poaching pressure on one of Malawi's premier tourism attractions.

## A fence, water and a fish farm

And so, discussions began between IFAW and its partners, and the people of Chikolongo. And it wasn't long before real, on the ground action was taking place.

In 2013 the Chikolongo Fish Farm and Fence Project, led by IFAW and working with the local community, together with Malawi's Department of National Parks and Wildlife (DNPW) and the German Embassy, opened to great excitement. Hundreds of people turned out while government and other dignitaries came from far and wide to celebrate an initiative intended to bring long-lasting economic independence and prosperity to the community: an initiative firmly aligned with the objectives of easing human-wildlife conflict and poaching pressure on the fish and other animals of the national park.

"This area has been a hot spot of human-wildlife conflict for a number of years" said Rachel Mazombwe-Zulu, Malawi's then Minister of Tourism, Wildlife and Culture.

> The Government of Malawi commends IFAW for implementing the project. We believe it to be an excellent way of addressing human-wildlife conflict along the boundary of the park. It has come at the right time when Government is emphasising a participatory approach to wildlife conservation.

As part of the initial phase of the programme IFAW, with additional funding assistance from GIZ (a German development agency), erected a four-and-a-half-mile-long electrified game-proof fence to create a physical and well-defined boundary between the Liwonde National Park and the Chikolongo community. It runs along the western perimeter of the park and is designed to prevent poachers from entering the park while deterring elephants from encroaching on farmland. The dividend was immediate and positive: crop yields improved because of the absence of elephants (in fact, there has only been one report of an elephant breaking through since the building of the fence and not a single human death); poaching became less frequent; and incidents of people being injured or killed in confrontations with elephants fell away. The fence is now maintained and patrolled by local community members, some of whom are former poachers.

The second immediate objective of the project was to provide easier access to fresh water without women and children having to run the gauntlet of hippos and crocodiles several times a day. To this end a diesel pump house was installed within the park with an underground pipeline running from it to a community tap station. IFAW also constructed water storage tanks in the village, all relatively simple interventions that not only benefit the immediate Chikolongo community but also serve neighbours and farmers further afield.

In 2015 Chikolongo's fresh water system was extended, and food production further improved. A solar pump system has since replaced the diesel pump allowing for additional community taps in community land adjacent to the original scheme. And a large communal storehouse was also built (Figure 17.2).

*Figure 17.2* The entrance to the Chikolongo fish farm, which will provide the community with clean water and a source of income (Riccardo Gangale © IFAW).

Historically, local farming had been restricted to maize and soybean cultivation, with animal protein coming from fish illegally taken from the Shire River. A more sustainable dietary source of protein was clearly needed and this became the main driver for an aquaculture alternative. And so the fish farm, originally constructed without any mechanised assistance by 24 families within the community, came into being. It spreads over an area more than half a mile long and some 550 yards wide, comprising seven ponds of varying sizes, one natural dam and one that has been built to support irrigation. The ponds were stocked with indigenous species naturally occurring in the Shire River and currently hold tilapia, fish well suited to aquaculture because of their high growth rate and excellent protein content.

## Honey, charcoal and tourism – an expanding bounty

Vegetables were planted along the pond edges and on the edges of the ponds. The fresh produce grown is sold to the nearby Mvuu Lodge in the park, enabling the Chikolongo villagers to benefit directly from the tourism industry, an opportunity that simply did not exist before the farm project.

Other opportunities have snowballed from the Chikolongo initiative. The safer, more productive environment has spawned a tree nursery, a growth in chicken, duck and goat husbandry, and beekeeping has taken root. Currently there are 40 hives, each one producing more than 20 pounds of honey every two months. At about £1.80 a pound it is yet another meaningful cash injection into the community coffers. And bees have another useful contribution to make – elephants have a natural aversion to them and will avoid coming close to their hives. If strategically placed the hives provide an effective second-tier defence against elephant raids.

The apiary now plays a core role in the sustainability of the Chikolongo project, but bees also have needs and if honey production is to reach its full potential a variety of pollen-bearing plant species and consistent supplies of water in order to thrive. Areas of indigenous trees and other plants are being planned in this regard.

This assessment of the bee-keeping project came as part of a thorough evaluation of the Chikolongo initiative commissioned by IFAW in 2017. The resultant report has highlighted the potential of a number of existing and new initiatives and how they could be developed. One of these is a bamboo nursery. Tough, fast-growing bamboos are extraordinarily versatile plants – more than 1,500 uses for their wood have been documented. The two indigenous species of bamboo found in Malawi are traditionally used to make fences for livestock and flat, plate-like baskets, but there is another important opportunity.

Dependence on wood charcoal is a major challenge for the communities surrounding Chikolongo Farm Project. In fact charcoal production is a key factor in Malawi's deforestation crisis. Growing bamboo, as well as other fast-growing indigenous trees, for this purpose could present an excellent alternative to tree cutting while also helping to combat soil erosion.

Additional opportunities highlighted in the evaluation include sustainable compost production, which would help return nutrients to the soil improve its water-holding ability. Already no chemical fertilisers and insecticides are used on the Chikolongo project farm. Other proposed projects include the building of a new conservation centre that would double as a community event venue and community-based tourism, offering cultural excursion packages to park visitors that would be informative, active, fun and a memorable experience. All could be developed without any significant new investment and all show the potential for Chikolongo as an innovation hub to test and pilot new ideas.

## Investing in children

In 2014 a unique conservation education partnership between IFAW and HELP Malawi (HELP stands for Help, Educate, Love, Protect) was launched with high hopes of improving school pass rates in the district and exposing children to conservation values. At the event Nancy Barr, Director of IFAW's Animal Action Education Programme, commented,

> This exciting initiative is developing locally-relevant education programmes intended to improve knowledge and attitudes among people, who share their lives daily with wildlife and, very importantly will improve graduation rates among the more than 1,000 school goers in this community and reduce the number of repeater students.

IFAW funding has also helped build two new dormitory blocks to house 12 trainee teachers, the presence of whom has reduced pupil/teacher ratios from 90:1 to 43:1 while providing the teaching assistance needed to integrate IFAW's conservation and animal welfare curriculum into school lessons.

The benefits are far reaching as it appears that the conservation of large animal species may depend more on good education, greater literacy, good government and less corruption than merely setting aside areas for conservation.

## A model for the future

There is no doubt that Chikolongo project has been a resounding success, predominantly due to the security provided by the fence and safe access to water. The standard of living has certainly improved for more the 6,000 individuals living in the community, who express overwhelming support for the project, which has helped to eliminate elephant attacks on themselves and their crops, and the threat from crocodiles and hippos when collecting water. This has especially helped women and girls, the traditional collectors of water and firewood. Not only has their personal safety improved dramatically, having water close to hand means more time to go to school.

Equally gratifying is the growing change in attitude towards wildlife – especially elephants. This is particularly manifest among children who see the connection between a healthy, thriving national park on their doorstep and their own well-being. "Wildlife is important to my community", they overwhelmingly agreed (IFAW research, unpublished).

Chikolongo has reached the point where the word is spreading about how such empowerment helps people and other communities are expressing their desire to be part of such a project.

Chikolongo has certainly delivered in terms of its wildlife objectives. In addition to the successes on the elephant front, there has been a measurable reduction in fish poaching in the Shire River. Poachers are abandoning their criminal ways and are becoming law-abiding farmers.

Most of all, the Chikolongo project's wide-ranging success is because we listened to the community and made the goal of animal protection with a community's needs in mind.

# 18 Silvopastoral systems

## A feasible step towards environmentally sustainable livestock systems in Mexico

*Karen F. Mancera*

### Challenges to the cattle industry in Mexico

The cattle industry in Mexico produces 1.8 million tonnes of beef per year, making it the seventh-largest beef producer of the world; as a consequence, livestock production is one of the major contributors to the country's gross domestic product (GDP) and one of the most important sources of employment (Heinrich-Böll-Stiftung et al., 2014). Beef production occurs at both large and small scales, with small-scale livestock operations in the country fostering better use of limited land resources and providing important improvements to farmers' incomes and diets (Wiggins et al., 2001; Arriaga-Jordán and Pearson, 2004). Such improvements are particularly important as small farm operations have proven to be significant contributors to reducing poverty in developing countries (Randolph et al., 2007). Despite its relative benefits, livestock production is also directly associated with deforestation, loss of biodiversity and a decrease in ecosystem services.

In Mexico, the principal cause of deforestation is the conversion of tropical dry forest into pastures with introduced grasses (Stern et al., 2002). The first step towards conversion to pasture is the burning of trees, thus eliminating ecosystem properties and services (Jaramillo et al., 2010). In addition, current livestock systems in Mexico have proven to be ineffective because of their low productivity and high incidence of disease (Amendola et al., 2005; FAO 2006; Murgueitio et al. 2011). Mexico has, therefore, committed to generating environmentally sustainable production through efficient practices in the use of natural resources and reducing environmental deterioration, as stated by the Mexican Secretariat of Agriculture, Livestock, Rural Development, Fisheries and Food (SAGARPA) in its National Farming Planning document (National Planning Document (SAGARPA, 2017). This stance has also been highlighted internationally as necessary to minimise the effects of livestock production by redirecting it to the provision of environmental services while maintaining food and economic security (FAO 2006; Balvanera et al., 2009). A feasible option to reach these objectives is the use of alternative livestock systems with a complex vegetation component.

## Silvopastoral systems as an option for livestock production

Silvopastoral systems (SPS) are associations of pasture, trees and/or shrubs. Although the implementation costs of an SPS can be high, they also increase economic returns for farmers, for instance, through additional revenues from timber sales (Solarte et al., 2011). SPS also improve farmers' quality of life by increasing job satisfaction and security (Broom et al., 2013). Furthermore, due to their vegetal complexity, these systems also provide environmental services, including better soil quality, carbon capture and higher biodiversity rates (Betancourt et al., 2003; Montagnini, 2009; Ibrahim et al., 2011; Broom et al., 2013).

In Mexico there is evidence of the unintentional use of SPS. For instance, in the state of Veracruz, the most important agroforestry system encountered was the SPS, and it was determined that the existence of trees in these systems was a product of natural succession, rather than an imposed design preconceived by farmers (Bautista Tolentino, 2009). In addition, farmers using SPS systems obtained different economic benefits derived from the arboreal component, such as firewood extraction and material for fences (Bautista Tolentino, 2009). SPS are also associated with the use of endemic plants and traditional breeds and technologies in Africa and South America (Gardiner and Devendra, 1995), which are factors often associated with sustainability.

Unfortunately, SPS in the Mexican tropics are poorly managed, which limits their total productive potential of the vegetal component, that is, the capacity of a plant to take full advantage of all the environmental factors available to promote its growth and performance in a determined geographic location (Villavicencio et al., 2007). However, the fact that SPS systems are present in the country because of traditional practices could facilitate their integration and expansion, as development strategies could focus on tools for improved management rather than having to build acceptance for the introduction of alternative systems. In recent years researchers have explored factors that make for successful alternative livestock systems in the Mexican tropics. This research has included the evaluation of cattle welfare in relation to SPS (Figure 18.1).

## SPS and animal welfare: an option for humane livestock production

Animal welfare refers to the state of an individual in relation to its ability to cope with its environment (Broom, 1991). Recently researchers have obtained evidence of improvements to animal well-being in the presence of trees. The research measured tree coverage and animal welfare indicators in ten cattle ranches based in the Mexican tropics. Researchers found that body condition was better and integument lesions were fewer in cattle found in ranches with tree coverage of more than 10%. Shorter flight distances (an indicator of reduced fear to humans) were also observed when cattle grazed in

*Figure 18.1* Silvopastoral systems have many benefits including for animal welfare and the environment.

*Source*: Pablo Pérez Lombardini and María Sarasvati Herrera.

wooded grassland (pastures associated with patches of trees) (Mancera et al., 2018). These results supported previous findings, in which increased body weight and health have been observed in cattle browsing leguminous trees often associated with SPS (Aguilar and Condit, 2001; Castañeda Nieto et al., 2003; Esquivel, 2007).

Research has also been conducted in Yucatán, Mexico, where behavioural and physiological welfare indicators have been measured and compared in SPS and monoculture systems (MS). It was found that mean skin temperatures were 1.21°C lower for cattle grazing in an SPS compared to an MS, even two hours after leaving the paddock for their night enclosures (Galindo et al., 2013), thus diminishing the chances of heat stress for cattle grazing in an SPS. Heifers also spent more time resting and presented more affiliative interactions (social licking, head leaning and social rubbing) (Améndola et al., 2016). In addition, when foraging behaviour (browsing and grazing) and the temperature humidity index (THI) were evaluated, foraging times in relation to THI were significantly decreased in the SPS compared to the MS, suggesting that the forage availability and access to shade in the SPS allow cattle to rest and ruminate longer, whereas cattle in an MS spend more time foraging at times of the day where the temperatures are higher (Amendola, in press), which could expose them to the detrimental effects of heat stress (Blackshaw and Blackshaw, 1994).

Overall, the improvements observed on animal welfare indicators in Mexican systems associated with trees indicate that SPS can decrease the stress caused by hunger, heat or fear. In return, the improvement of animal welfare in SPS can generate higher weight gains and animal products containing higher protein and lower fat content (Corral et al., 2011; Paciullo et al., 2011). The provision of animal products of better quality sourced from environmentally conscious and compassionate systems is becoming increasingly relevant in Mexico, as several sectors of the public are demanding positive developments in the livestock industry (Miranda-de la Lama et al., 2017).

Finally, the introduction of animal welfare improvements in relation to the implementation and successful management of SPS is a definitive way to develop humane livestock practices. Humane livestock systems involve the use of traditional small-scale farms that deliver animal products obtained ethically for the benefit of consumers, producers, animals and the environment (D'Silva, 2013). Although meat consumption should be reduced to achieve ethical goals (D'Silva, 2013), humane and sustainable livestock production systems have a role to play when we consider the projected population growth of 8.6 billion by 2030 (with 653 million malnourished; United Nations, 2017), the expected food yield increase of 80% of the existing crop land in 2050 (thus increasing land pressure and chances of degradation; FAO, 2014), and the fact that 73% of livestock production occurs in arid land (FAO, 2012) that is likely unsuitable for crop production. Ruminants in SPS and other pasture-based systems, in particular, make an important contribution to food security by converting food that humans cannot eat into food that they can.

## Conclusion

Developing countries need to find livestock practices that maintain food security while preserving environmental services and sustainability. SPS are a good option in Mexico as they are related to traditional practices and promote good welfare for the animals reared. Conversion to such systems generates additional revenues and economic sustainability for farmers through correct management and commercialisation of the vegetal component as well as increased quality of products. Furthermore, SPS promote higher ethical standards that benefit animals, producers, consumers and the environment in a global scenario in which livestock production should play its role in supporting better land use and increasing food security.

## Acknowledgements

This contribution is part of Karen F. Mancera postdoctoral project, supported by the postdoctoral scholarship programme UNAM-DGAPA at the Faculty of Veterinary Medicine – UNAM.

## References


Aguilar, S. and Condit, R. 2001. Use of native tree species by an Hispanic community in Panama. *Economic Botany,* 55, 223–235.

Améndola, L., Solorio, F., Ku-Vera, J., Améndola-Massiotti, R., Zarza, H. and Galindo, F. 2016. Social behaviour of cattle in tropical silvopastoral and monoculture systems. *Animal,* 10, 863–867.

Amendola, L., Solorio, F.J., Ku – Vera, J.C., Amendola – Massioti, R.D., Zarza, H., Mancera, K.F., and Galindo, F. A pilot study on the foraging behaviour of heifers in intensive silvopastoral and monoculture systems in the tropics. *Animal,* in press.

Améndola, R., Castillo, E. and Arturo, P. 2005. *Perfiles por país del recurso pastura/forraje.* México II. Available at www.fao.org/ag/AGP/AGPC/doc/Counprof/spanishtrad/Mexico_sp/Mexico_sp.htm. (Accessed on 23 April 2018).

Arriaga-Jordán, C.M. and Pearson, R. 2004. The contribution of livestock to smallholder livelihoods: the situation in Mexico. *BSAS Occasional Publication,* 33, 99–116.

Balvanera, P., Cotler, H., Aburto, O., Aguilar, A., Aguilera, M., Aluja, M., Andrade, A., Arroyo, I., Ashworth, L., Astier, M. and Ávila, P. 2009. Estado y tendencias de los servicios ecosistémicos, in Dirzo, R., González, R. and March, I.J. (eds) *Capital Natural de México, vol. II: Estado de conservación y tendencias de cambio.* Comisión Nacional para el Conocimiento y Uso de la Biodiversidad, Mexico, pp. 185–54.

Bautista Tolentino, M. 2009. *Sistemas agro y silvopastoriles en El Limón, municipio de Paso de Ovejas, Veracruz, México.* Master's thesis. Colegio de Posgraduados, Campus Veracruz.

Betancourt, K., Ibrahim, M., Harvey, C. and Vargas, B. 2003. Efecto de la cobertura arbórea sobre el comportamiento animal en fincas ganaderas de doble propósito en Matiguás, Matagalpa, Nicaragua. *Agroforestería en las Américas,* 10, 47–51.

Blackshaw, J.K. and Blackshaw, A. 1994. Heat stress in cattle and the effect of shade on production and behaviour: a review. *Animal Production Science,* 34, 285–295.

Broom, D., Galindo, F. and Murgueitio, E. 2013. Sustainable, efficient livestock production with high biodiversity and good welfare for animals. *Proceedings of the Royal Society B: Biological Sciences,* 280(1771), 20132025.

Broom, D.M. 1991. Animal welfare: concepts and measurement. *Journal of Animal Science,* 69, 4167–4175.

Castañeda Nieto, Y., Álvarez Morales, G. and Melgarejo Velazquez, L. 2003. Ganancia de peso, conversión y eficiencia alimentaria en ovinos alimentados con fruto (semilla con vaina) de parota (*Enterolobium cyclocarpum*) y pollinaza. *Veterinaria México OA,* 34.

Corral, G., Solorio, B., Rodríguez, C. and Ramírez, J. 2011. La calidad de la carne producida en el sistema silvopastoril intensivo y su diferenciación en el mercado, in *Memorias III Congreso sobre Sistemas Silvopastoriles Intensivos, para la ganadería sostenible del siglo XXI,* Fundación Produce Michoacán, COFRUPO, SAGARPA, Universidad Autónoma de Yucatán–UADY, Morelia, México, pp. 46–52.

D'Silva, J. 2013. The meat crisis: the ethical dimensions of animal welfare, climate change, and future sustainability, in *Sustainable food security in the era of local and global environmental change,* Springer, Dordrecht, pp. 19–32.

Esquivel, H. 2007. *Tree resources in traditional silvopastoral systems and their impact on productivity and nutritive value of pastures in the dry tropics of Costa Rica.* PhD thesis, CATIE.

FAO. 2006. *Livestock's long shadow. Environmental issues and options*. The Food and Agriculture Organization of the United Nations, Italy, Rome.

FAO. 2012. *Livestock and landscapes*, Italy, Rome.

FAO. 2014. *Building a common vision for sustainable food and agriculture*, Italy, Rome.

Galindo, F., Olea, R., and Suzan, G. 2013. *Animal welfare and sustainability*. International Workshop on Farm Animal Welfare. Sao Paulo, Brazil.

Gardiner, P. and Devendra, C. 1995. *Global agenda for livestock research: proceedings of a consultation, 18–20 January 1995*, ILRI (aka ILCA and ILRAD), Nairobi, Kenya.

Heinrich-Böll-Stiftung, Chemnitz, C. and Becheva, S. 2014. *Meat atlas: facts and figures about the animals we eat (Spanish version)*, Heinrich Böll Foundation, Germany.

Ibrahim, M., Casasola, F., Villanueva, C., Murgueitio, E., Ramírez, E., Sáenz, J. and Sepúlveda, C. 2011. Payment for environmental services as a tool to encourage the adoption of silvopastoral systems and restoration of agricultural landscapes dominated by cattle in Latin America. In *Restoring Degraded Landscapes in Latin America*, Nova Science Pub Inc, New Haven, CT, pp. 1–23.

Jaramillo, V., García-Oliva, F. and Martínez-Yrízar, A. 2010. La selva seca y las perturbaciones antrópicas en un contexto funcional, in Ceballos, G., Martínez, L., García, A., Espinoza, E., Creel, J.B. and Dirzo, R (eds) *Diversidad, amenazas y áreas prioritarias para la conservación de las selvas secas del Pacífico de México*, FCE-CONABIO-TELMEX-CONANP-WWF México-EcoCiencia SC, Mexico, pp. 235–250.

Mancera, K.F., Zarza, H., De Buen, L.L., García, A.A.C., Palacios, F.M. and Galindo, F. 2018. Integrating links between tree coverage and cattle welfare in silvopastoral systems evaluation. *Agronomy for Sustainable Development*, 38, 19.

Miranda-De La Lama, G., Estévez-Moreno, L., Sepúlveda, W., Estrada-Chavero, M., Rayas-Amor, A., Villarroel, M. and María, G. 2017. Mexican consumers' perceptions and attitudes towards farm animal welfare and willingness to pay for welfare friendly meat products. *Meat Science*, 125, 106–113.

Montagnini, F. 2009. El pago de servicios ambientales (PSA) como herramienta para fomentar la restauración y el desarrollo rural. Paper presented at Congreso Forestal Mundial. Buenos Aires, Argentina.

Murgueitio, E., Calle, Z., Uribe, F., Calle, A. and Solorio, B. 2011. Native trees and shrubs for the productive rehabilitation of tropical cattle ranching lands. *Forest Ecology Management*, 261, 1654–1663.

Paciullo, D.S.C., De Castro, C.R.T., De Miranda Gomide, C.A., Maurício, R.M., Pires, M.D.F.Á., Müller, M.D. and Xavier, D.F. 2011. Performance of dairy heifers in a silvopastoral system. *Livestock Science*, 141, 166–172.

Randolph, T., Schelling, E., Grace, D., Nicholson, C.F., Leroy, J., Cole, D., Demment, M., Omore, A., Zinsstag, J. and Ruel, M. 2007. Invited review: role of livestock in human nutrition and health for poverty reduction in developing countries. *Journal of Animal Science*, 85, 2788–2800.

SAGARPA. 2009. *Escenarios base 09-18*. Proyecciones para el Sector Agropecuario, México, D.F.

SAGARPA. 2017. *Planeación Agrícola Nacional 2017–2030*, México, D.F.

Solarte, L., Cuartas, C., Naranjo, J., Uribe, F. and Murgueitio, E. 2011. Estimación de los costos de establecimiento para sistemas silvopastoriles intensivos con *Leucaena leucocephala*, pasturas mejoradas y árboles maderables en el Caribe seco Colombiano. *Revista Colombiana de Ciencias Pecuarias*, 24, 518.

Stern, M., Quesada, M. and Stoner, K.E. 2002. Changes in composition and structure of a tropical dry forest following intermittent cattle grazing. *Revista de Biología Tropical,* 50, 1021–1034.

United Nations. 2017. www.un.org/development/desa/publications/world-population-prospects-the-2017-revision.html. (Accessed on 23 April 2018).

Villavicencio, E., Martínez, O., Cano, A. and Berlanga, C. 2007. Orégano, recurso con alto potencial. *Ciencia y Desarrollo,* 33, 60–66.

Wiggins, S., Tzintzun-Rascón, R., Ramírez-González, M., Ramírez-González, R., Ramírez-Valencia, F.J., Ortiz-Ortiz, G., Piña-Cárdenas, B., Aguilar-Barradas, U., Espinoza-Ortega, A., Pedraza-Fuentes, A.M., Rivera-Herrejón, G. and Arriaga-Jordán, C.M. 2001. *Costos y Retornos de la Producción de Leche en Pequeña Escala en la Zona Central de México. La lechería como empresa.* Serie Cuadernos de Investigación. Cuarta Época 19. Universidad Autónoma del Estado de México, Toluca, México.

# 19 Farming in harmony with nature

*Patrick Holden*

## What is the problem with food and farming?

It has become a cliché but it's true: supermarket food is not really cheap; it comes at a heavy hidden cost for which we pay in hidden ways. The industrial application of nitrogen fertiliser has contaminated our water systems and atmosphere with dangerous nitrates; the subsidised production of fructose corn syrup has driven an increase in obesity and diabetes; and the excessive use of antibiotics in animals has caused a resistance to these drugs amongst humans.

In a world where intensive farming, especially that of livestock, has become the most profitable option for producing food, the result has been a physical disconnection from the natural world by both farmers and consumers. A large proportion of our food is now grown in systems where nature and agriculture are completely separate. Soil is treated as an inert medium with only chemical and physical properties, instead of vibrant biological life, the home of a quarter of the planet's biodiversity. Synthetic pesticides are needed on crops because pests and predators are out of harmony; synthetically modified feeds, antibiotics and wormers are needed for livestock because we no longer allow animals to graze naturally in the fields and develop strong immune systems.

This reductionist, separationist approach to farming has made it difficult to argue an economic case for sustainable farming practice. As an organic dairy farmer, the reason I need to charge more for the food I produce is because I am "internalising" costs which other producers pass on to the environment, future generations and government departments, not least of which is the Department of Health. This occurs because farmers do not have to pay for diffuse pollution and degradation of the air, water or soil, or for the impact on human health of food with low nutritional quality or contamination. A report by the Sustainable Food Trust, *The Hidden Cost of UK Food,* published in November 2017, identifies the hidden costs associated with current intensive modes of farming in the UK, which range from diet-related ill-health, food contamination and reduced quality, antimicrobial resistance, air, soil and water pollution, to soil degradation, biodiversity loss, global warming and the depletion of natural resources. The primary purpose of the report is to estimate the total value of the negative impacts and the potential benefits for society if these costs were to be reduced.

Breaking down the annual costs, the report finds that in addition to the £120 billion spent annually on food by consumers, the UK food system generates further costs of £120 billion, nearly 30 times higher than previous composite estimates have indicated. These extra costs are not paid by the food businesses that cause them, nor are they included within the retail price of food. Instead they are passed on to society in a range of hidden ways – meaning that UK consumers are, in effect, paying twice for their food. It is striking to note that a system with a shadow economy of over £120 billion a year, self-identifies as "efficient" (Sustainable Food Trust, 2017).

The current policy framework supports a dishonest economic food pricing system, as a result of which, the best business case is for farmers to grow using industrial methods and for retailers to buy the commodity products from industrial farms, process and package them so the consumer knows nothing about their backstory and then make a profit by turning that around. So, we need new incentives and disincentives, which ensure that the polluter pays and those who farm in a truly sustainable way are better rewarded for the benefits they deliver.

## A healthy and sustainable future

Despite the visible separation affecting our current thinking, I have always believed that it is possible, if we farm in harmony with nature, for the vast community of bacteria, fungi, soil-invertebrates, wild flowers, insects, amphibians, birds and small mammals to coexist in a balanced way with a food production system – even if the farmer, in this case me, is determined to produce as much food as possible.

Britain's conservation organisations recently acknowledged in their *State of Nature* report that despite their best intentions to design stewardship schemes to protect our wild plants and animals, there has been a relentless decline in all the key indicator species over the last 50 years (Hayhow, Burns, Eaton et al., 2016). All this raises the question of whether there is a better way of encouraging an abundance of wildlife on Britain's farms? And that is where direct observations of my own farm, 40 years after its establishment, can be drawn upon.

Regrettably my own farm in Calon Wen in Wales has not been subject to biodiversity audits over the last four decades; thus my observations are not based upon good science, and not peer-reviewed. Nevertheless, they provide at least anecdotal evidence of the point I'm trying to get across. Come and stand in our yard and listen to the incredible abundance of bird song, or take a plunge, as I do most days, in our spring-fed farm pond and witness the incredible diversity of nature, including hundreds of freshly metamorphosed toads and mating dragonflies crawling up the banks, all flourishing in the very heart of our farm. It's not that we've got every species in the book, but I'm certain that we have much higher numbers of birds, such as house sparrows, starlings, chaffinches, owls, tits, swallows and birds of prey that feed on the lower links in the food chain, including the countless species of insects (which gravitate to our cow pats) which aren't, of course, killed by wormers such as Ivermectin, and which proliferate and coexist where nitrogen fertiliser and pesticides are not used.

Why is this the case? Because if you farm in harmony with nature, without the use of chemical inputs, each field becomes a food source, rather than a monoculture of ryegrass, or wheat, or oilseed rape, where virtually nothing lives except the cultivated plant.

There are other changes needed to ensure a sustainable, and economically viable future for farmers in the UK. A return to mixed farming, involving a fertility-building phase, primarily of grass and legumes, accompanied by grazing ruminants, is part of the solution. In terms of addressing climate change and nutritional quality, contrary perhaps to widespread public opinion, grass-fed ruminants are nutritionally and environmentally part of the solution. We must avoid throwing out the grass-fed ruminant baby with the bathwater of feedlot beef and industrial livestock production. Instead, it is important that we are able to differentiate between livestock systems that are part of the solution and those that are part of the problem.

I strongly believe that a healthy diet should work backwards from the most sustainable way to farm. The big challenge is that a systemic switch is needed to sustainable food production. The landscape created by such a change will necessitate soil fertility-building grassland, crop rotations and ruminant grazing. In order to make this system economically viable, we need to match our diets with the livestock products derived from such sustainable systems. The phrase "eat less meat but better" is often used, but what this really means is that we, as consumers, must give up cheap, industrially farmed livestock such as chicken and pork, as well as cutting out industrially produced dairy products. It also means transferring your livestock product loyalty to grass-fed ruminant meat, pasture-fed chicken and pigs fed on food waste.

For too long we have dined out on the natural capital of the soil that previous generations have laid down for us. We need to fix that, not only for the future of a healthy planet, but because the environment in which a plant or animal is produced goes a long way to determine its nutrient value when consumed by humans.

## References

Fitzpatrick I, Young R et al. 2017, The Hidden Cost of UK Food, Sustainable Food Trust.

Hayhow DB, Burns F, Eaton MA, Al Fulaij N, August TA, Babey L, Bacon L, Bingham C, Boswell J, Boughey KL, Brereton T, Brookman E, Brooks DR, Bullock DJ, Burke O, Collis M, Corbet L, Cornish N, De Massimi S, Densham J, Dunn E, Elliott S, Gent T, Godber J, Hamilton S, Havery S, Hawkins S, Henney J, Holmes K, Hutchinson N, Isaac NJB, Johns D, Macadam CR, Mathews F, Nicolet P, Noble DG, Outhwaite CL, Powney GD, Richardson P, Roy DB, Sims D, Smart S, Stevenson K, Stroud RA, Walker KJ, Webb JR, Webb TJ, Wynde R and Gregory RD 2016, *State of Nature 2016*. The State of Nature partnership.

# Part VI

# Food policy for the future

# 20 Paying for the true costs of our meat, eggs and dairy

*Peter Stevenson*

The bleak lives imposed on industrially farmed animals are justified by the assertion that this gives us cheap food. But the low cost of animal products is achieved only by an economic sleight of hand. We have devised a distorting economics which takes account of some costs such as housing and feeding animals but ignores others including the detrimental impact of industrial agriculture on human health and natural resources.

Industrial livestock production contributes to impaired human health, overuse of antimicrobials, environmental degradation, greenhouse gas (GHG) emissions, loss of biodiversity and wildlife and very poor animal welfare.

These various detrimental impacts are referred to by economists as "negative externalities". They represent a market failure in that the costs associated with them are borne by third parties or society as a whole and are not included in the costs paid by farmers or the prices paid by consumers of livestock products. In some cases the costs are borne by no-one and key resources such as soil and biodiversity are allowed to deteriorate, undermining the ability of future generations to feed themselves. When such externalities are not included in prices, they distort the market by encouraging activities that are costly to society even if the private benefits are substantial (Petty et al., 2001). The failure to reflect full social costs arguably leads to private gains being viewed as more important than public losses.

## Need to internalise externalities is widely recognised

There is increasing recognition that, in order to avoid market distortions and encourage efficient use of scarce resources, these externalities should be internalised in the costs of producing meat, milk and eggs and thus in the price paid by consumers.

A report by the UN Food and Agriculture Organisation (FAO) has said, "In many countries there is a worrying disconnect between the retail price of food and the true cost of its production. As a consequence, food produced at great environmental cost in the form of greenhouse gas emissions, water pollution, air pollution, and habitat destruction, can appear to be cheaper than more sustainably produced alternatives" (FAO, 2015).

The Foresight report on the future of food and farming said, "There needs to be much greater realisation that market failures exist in the food system that, if not corrected, will lead to irreversible environmental damage and long term threats to the viability of the food system. Moves to internalise the costs of these negative environmental externalities are critical to provide incentives for their reduction" (Foresight, 2011).

## The negative externalities of industrial livestock production

*The negative health externalities* of industrial livestock production arise from several factors:

- The high levels of meat consumption that have been made possible by industrial farming can lead to heart diseases, obesity, diabetes and certain cancers (European Commission, 2012).
- Industrially farmed animals are routinely given antibiotics to ward off the diseases that would otherwise be inevitable when large numbers of animals are kept in crowded, stressful conditions.
- Air pollution arising from agriculture.
- Exposure to agro-chemicals.
- Food-borne diseases such as salmonella and campylobacter.

*The negative environmental externalities* of industrial livestock production largely stem from its dependence on feeding human-edible cereals to animals. This is inherently inefficient as much of the food value of crops is lost during conversion from plant to animal matter. Studies show that for every 100 calories of cereals fed to animals, just 17–30 calories are delivered to the human food chain as meat (Lundqvist et al., 2008; Nellemann et al., 2009). Indeed the efficiency rates may be even lower for some animal products (Cassidy et al., 2013). The FAO warns that further use of cereals as animal feed could threaten food security by reducing the grain available for human consumption (Gerber et al., 2013).

Industrial livestock's huge demand for cereals has fuelled the intensification of crop production. This, with its monocultures and agro-chemicals, has led to soil degradation, overuse and pollution of ground-and surface water, biodiversity loss and air pollution. The relentless need for cereals and soy as animal feed is also driving expansion of arable land into grasslands, savannahs and forests. This leads to loss and fragmentation of wildlife habitats and the release of stored carbon into the atmosphere.

## Estimating the costs entailed in negative externalities

### *Excess nitrogen in the environment*

The *European Nitrogen Assessment* (ENA) reports that 75% of industrial production of reactive nitrogen ($N_r$) in Europe is used for fertiliser (2008 figure).

The ENA points out that the primary use of $N_r$ in crops in Europe is not directly to feed people but to provide feeds to support livestock.

The ENA identifies five key threats associated with excess $N_r$ in the environment: damage to water quality, air quality (and hence human health, in particular respiratory problems and cancers), soil quality (acidification of agricultural soils and loss of soil biodiversity), the greenhouse balance and ecosystems and biodiversity (Sutton et al., 2011).

The ENA points out that although the atmospheric emissions of nitrogen oxide from traffic and industry contribute to many environmental effects, these emissions are dwarfed by the agricultural flows of reactive nitrogen.

The ENA estimates that the cost of environmental damage related to $N_r$ effects from agriculture in the EU-27 is €20–€150 billion per year. A cost-benefit analysis shows that this outweighs the benefit of N-fertiliser for farmers of €10–€100 billion per year.

### *Soil degradation*

A UK study concludes that "modern agriculture, in seeking to maximize yields...has caused loss of soil organic carbon and compaction, impairing critical regulating and supporting ecosystem services" (Edmondson et al., 2014). The study point outs that depletion of soil organic carbon "in conventional agricultural fields is now thought to be an important factor constraining productivity as many arable soils have suboptimal concentrations".

The consequences of soil biodiversity mismanagement have been estimated to be in excess of one trillion dollars per year worldwide (Turbé et al., 2010). The FAO stresses, "the current rate of soil degradation threatens the capacity to meet the needs of future generations, unless we reverse this trend through a concerted effort towards the sustainable management of soils" (FAO, 2014a). The FAO estimates that worldwide 75 billion tonnes of soil are lost every year, costing approximately US$400 billion per year (Steinfeld et al., 2006).

### *Overuse and pollution of water*

Industrial livestock production generally uses and pollutes more groundwater and surface water than grazing or mixed systems (Mekonnen and Hoekstra, 2012). Unabsorbed nitrogen and phosphorus from fertilisers and chemicals from pesticides are key factors in water pollution. The Organisation for Economic Co-operation and Development (OECD) reports that the annual costs related to water pollution in six European Union (EU) Member States (Belgium, France, the Netherlands, Sweden, Spain, and the UK) amount to €2.43–€4.75 billion per year (Moxey, 2012).

A US study found that in all US nutrient ecoregions nitrogen and phosphorus concentrations in rivers and lakes exceeded reference values (Dodds et al., 2009). In 12 out of 14 ecoregions, over 90% of rivers exceeded reference values. The study calculated potential annual value losses in drinking water treatment costs, recreational water usage, spending on recovery of threatened and endangered species, and waterfront real estate. The combined costs were approximately $2.2 billion annually as a result of eutrophication in US freshwaters. The study recognises that a substantial portion of human-induced eutrophication ultimately stems from fertiliser use. The authors point out that their evaluation likely underestimates the economic losses.

### *Biodiversity loss*

The European Environment Agency estimates that biodiversity loss reduces global gross domestic product (GDP) by 3% each year (EEA, 2015). Globally food production accounts for 60%–70% of total biodiversity loss (Kok et al., 2014). The European Commission states that the livestock sector may be the leading player in the reduction of global biodiversity through its demand on land (European Commission, 2011a).

Intensive agriculture has also played a major role in the decline in pollinators such as bees through its use of insecticides and herbicides and its contribution to air pollution and habitat deterioration (Kluser et al., 2010; Wentworth, 2013). The value of insect pollination services to crop agriculture has been estimated at €153 billion globally (Kluser et al., 2010).

### *Climate change*

The FAO estimates that the livestock sector is responsible for 14.5% of human-induced GHG emissions (Gerber et al., 2013). Studies suggest that "business-as-usual" (BAU) will lead to agriculture's GHG emissions being so high by 2050 that they alone will push global temperatures to increase by almost 2°C (Bailey et al., 2014; Bajželj et al., 2014).

Springmann et al. (2016a) compared the health impacts in 2050 of a reference diet based on FAO projections with three alternatives: (i) a healthy global diet based on WHO/FAO Expert Consultations and recommendations by the World Cancer Research Fund, (ii) a vegetarian diet and (iii) a vegan diet (Springmann et al., 2016a) The researchers estimate that, compared with the reference diet, the adoption of a healthy global diet would have monetised environmental benefits due to reduced GHG emissions of $234 billion per year. Adoption of the vegetarian and vegan diets would have benefits, compared with the reference diet, of $511 and $570 billion per year, respectively.

## The environmental costs of feeding cereals to animals

Globally 36% of cereal production is used as animal feed (Cassidy et al., 2013). As indicated earlier, animals convert cereals very inefficiently into meat and milk. This is a wasteful use not just of these crops but of the land, water and energy used to produce them. In addition, growing these cereals (most of which are produced intensively) for animal feed entails detrimental impacts such as water and air pollution, soil degradation, biodiversity loss and GHG emissions. As most of the calories and protein contained in the cereals are not converted into meat or milk, a substantial proportion of these detrimental impacts are produced for no purpose in terms of human food supply.

Research funded by the FAO calculates the difference in environmental impacts in 2050 between (i) BAU as regards feeding human-edible crops to animals and (ii) ending the use of such crops as animal feed (Schader et al., 2015). This will result in reduced production of meat and milk; the researchers have taken account of the fact that this will necessitate increased production of other foods but these generally will be much less resource-intensive. Table 20.1 shows the environmental impacts and the associated costs in 2050 of continuing with BAU as compared with ending the use of human-edible crops as feed.

The costs in the final column of Table 20.1 have been calculated using the data in Table 20.2 and elsewhere in the FAO report *Food waste footprint: full*

*Table 20.1* Comparison of impacts and their associated costs in 2050 between (i) BAU in use of human-edible crops as animal feed and (ii) ending use of such crops as feed*

| *Production inputs and environmental outcomes* | *Reference scenario: FAO BAU projections for 2050* | *No use of human-edible crops as animal feed in 2050* | *Extra impact of BAU compared with ending use of human-edible crops as animal feed in 2050* | *Extra cost of BAU compared with ending use of human-edible crops as animal feed in 2050* |
|---|---|---|---|---|
| Arable land use: million hectares | 1630 | 1200 | 430 | No figure available |
| GHG emissions: Gt CO2-eq | 12.8 | 10.4 | 2.4 | $271 billion per year |
| Freshwater use (for irrigation): km$^3$ | 2178 | 1718 | 460 | $575 billion per year[1] |
| N-surplus: million tonnes N | 121.8 | 65.2 | 56.6 | $17.5 billion per year[2] |
| P-surplus: million tonnes P | 64.0 | 38.4 | 25.6 | $322 billion per year[2] |
| Non-renewable energy use: exajoules | 26.7 | 17.2 | 9.5 | No figure available |
| Pesticide use:** | 15.4 | 12.0 | 3.4 | $38 billion per year[3] |
| Deforestation: million hectares | 7.2 | 6.5 | 0.7 | $1.1 billion per year[4] |
| Soil erosion from water: billion tonnes soil lost | 36.8 | 32.2 | 4.6 | $99 billion per year[5] |

1 This figure includes the cost of water use and the impact on water scarcity.
2 This figure includes eutrophication impact on both water quality and biodiversity.
3 This figure relates to health effects due to pesticide exposure.
4 This figure relates to loss of ecosystem services from deforestation; the cost of GHG emissions from deforestation is included in the figure for GHG emissions.
5 This figure relates to damage costs on-site and off-site.
* Costs are in US dollars (2012).
** Classification of pesticide use per hectare by intensity and crop, legislation by country and access to pesticides by farmers.

*cost accounting* (FAO, 2014b). The Annex to this paper details how the costs in the final column of Table 20.1 have been calculated.

Table 20.1 of this paper shows that the projected BAU use of human-edible crops as feed in 2050 will entail costs of $1323 billion (i.e. $1.32 trillion) per year as compared with not using such crops as animal feed. These costs arise mainly due to the inefficiency with which animals convert crops into meat and milk. The calculations presented in Table 20.1 are inevitably estimates. However, the overall cost may be much greater than $1.32 trillion per year as

- Data was not available to enable this paper to estimate the costs of arable land and energy use.
- This FAO report on food waste also included costs in respect of pollinator loss, the impact of pesticides on biodiversity, and loss of livelihoods and increased risk of conflict due to soil erosion. These aspects have not been included in this paper when estimating the costs arising from the projected BAU use of human-edible crops as feed in 2050 as the data to make reliable calculations is not available.

## Human health externalities

### *Non-communicable disease*

The high levels of meat consumption that have been made possible in the Western world by industrial farming are having an adverse impact on human health (Aston et al., 2012; European Commission, 2012; World Cancer Research Fund and American Institute for Cancer Research, 2011).

As indicted earlier, Springmann et al. (2016a) compared the health impacts in 2050 of a reference diet based on FAO projections with three alternatives: a healthy global diet based on WHO/FAO Expert Consultations and recommendations by the World Cancer Research Fund, a vegetarian diet and a vegan diet (Springmann et al., 2016a). The researchers estimate that adopting the healthy global diet rather than the reference diet would produce health-related cost savings of $735 billion per year. Adoption of the vegetarian and vegan diets would result, respectively, in savings of $973 and $1,067 per year. These benefits arise principally from reduced consumption of red meat, increased consumption of fruit and vegetables and limiting excessive energy intake. These figures are calculated using a cost-of-illness approach. A value-of-statistical-life approach led to much higher estimates of the economic benefits associated with dietary change.

### *Antimicrobial resistance*

The use of antimicrobials in human medicine is the main driver of antimicrobial resistance. However, the WHO has stressed that over-reliance on antimicrobials in intensive livestock farming is also a significant contributor to the emergence of antimicrobial-resistant bacteria that affect human health (WHO, 2012).

Each year 25,000 patients die in the EU from an infection caused by resistant microorganisms with extra healthcare costs and productivity losses of at least €1.5 billion per year (European Commission, 2011b). A study commissioned by the UK Government shows that a continued rise in resistance by 2050 would lead to 10 million more people dying worldwide every year than would be the case if resistance was kept to today's level and a reduction of 2%–3.5% in GDP (O'Neill, 2014). The study estimates that between now and 2050 the world can expect to lose between 60 and 100 trillion USD worth of economic output if antimicrobial drug resistance is not tackled.

The US Centers for Disease Control and Prevention (CDC) estimates that at least 2 million people are infected with antibiotic-resistant bacteria every year in the US with at least 23,000 people dying every year as a direct result of these infections (CDC, 2013). This incurs annual treatment costs of around $20 billion on top of costs to society for lost productivity that are as high as $35 billion a year, totalling $55 billion per annum (CDC, 2013).

### *Air pollution*

Agriculture is a key source of three major air pollutants: ammonia, particulate matter and nitrous oxide. Air pollution is a serious problem for human health as it contributes to conditions such as bronchitis, asthma, lung cancer and congestive heart failure. The related costs are considerable. A study has analysed the impact of Danish emission sectors on health-related costs arising from air pollution in Europe (Brandt, 2011). Emissions in Denmark cause health-related costs in Europe of €4.9 billion per year. The study found that agriculture is the main Danish sector contributing to health-related costs arising from air pollution in Europe; agriculture's contribution (43%) outweighs those of road traffic (18%) and major power plants (10%). A study for the US suggests that a 10% reduction in livestock ammonia emissions can lead to over $4 billion annually in particulate-related health benefits (Mccubbin et al., 2002).

A 2015 report by the French Senate concludes that the total cost of air pollution in France is between €68 and €97 billion per year (Aichi, 2015). This includes the medical costs of treating ill health resulting from air pollution such as certain cancers, asthma, bronchitis and cardiovascular problems. It also includes the costs of lost production as well as placing an economic value on loss of life and years of life spent in poor health. The study states that air pollution is mainly caused by four sectors: agriculture, transport, industry and residential. It does not provide an indication of the proportion of overall costs attributable to agriculture.

### *Exposure to agro-chemicals*

Recent research explores the health impacts of pesticides as "endocrine disrupting chemicals" (chemicals that interfere with hormones). A report by the TEEB for Agriculture and Food states that in the EU, of all endocrine disrupting chemicals, "pesticide exposure causes the highest annual health and economic costs at roughly $127 billion, almost four times as high as the second highest category (plastics)" (TEEB, 2015; Trasande et al., 2015).

### *Foodborne disease*

A US study estimates that the cost of foodborne illness in the US is $152 billion a year. This figure includes medical costs (hospital services, physician services and drugs) and quality-of-life losses (deaths, pain, suffering and functional disability) (Scharff, 2010).

A University of Florida study ranked the top 10 pathogen-food combinations and concluded that campylobacter in poultry was the most damaging in terms of both cost of illness and loss of Quality Adjusted Life Years (QALYs), a measure of health-related quality of life (Batz et al., 2011). Salmonella in poultry was the fourth most damaging. The study found that contaminated poultry has the greatest public health impact among foods. It is responsible for over $2.4 billion in estimated costs of illness annually and loss of 15,000 QALYs a year. Nearly all US chickens are produced industrially.

A 2015 study by the US Department of Agriculture estimates that foodborne illnesses impose over $15.5 billion in economic burden annually in the US (Hoffman et al., 2015).

#### *Campylobacter*

The European Food Safety Authority (EFSA) estimates that there are approximately nine million cases of human campylobacteriosis per year in the EU27. The disease burden of campylobacteriosis and its sequelae in the EU is 0.35 million disability adjusted life years per year and total annual costs are € 2.4 billion (EFSA BIOHAZ Panel, 2011). Poultry is a major source of campylobacters (Lyne et al., 2007).

#### *Salmonella*

EFSA states that over 100,000 human cases of salmonellosis are reported each year in the EU. EFSA has estimated that the overall economic burden of human salmonellosis could be as high as €3 billion a year (EFSA, no date). EFSA points out that salmonella is most frequently found in eggs and raw meat from pigs, turkeys and chickens (EFSA, no date). Most poultry and pig production in the EU is industrial.

## Animal welfare

Industrial livestock production generally results in low standards of animal welfare. A Dutch study seeks to quantify and value the adverse impact of pork production on pig welfare (Van Drunen et al., 2010). Based on willingness-to-pay research, the Dutch study suggests that the animal welfare-related costs of producing 1 kg of fresh pork are between €1.10 and €4.60 for conventionally produced pork and between €0 and €3.50 for organic pork. Taking the lower of these figures for conventionally produced pork and assuming that at least 90% of EU pigs are farmed intensively, the animal welfare costs of the EU pig sector are €19 billion per year.

All the above costs are set out in Table 20.2.

*Table 20.2* Costs of negative externalities

| *Negative externality* | *Estimated cost* | *Source* |
|---|---|---|
| UK impacts attributable to farming: GHG emissions, water pollution, air pollution, habitat destruction, soil erosion and flooding | £700 million per year | Natural Capital Committee |
| EU: environmental damage related to Nitrogen effects from agriculture | €20–€150 billion per year | European Nitrogen Assessment |
| EU: soil degradation | €38 billion per year | European Commission |
| Global: soil loss | $400 billion per year | FAO |
| UK: cleaning nitrates, pesticides, etc. from water | £271 million per year | O'Neill |
| US: eutrophication in freshwaters | $2.2 billion per year | Dodds et al |
| Belgium, France, the Netherlands, Sweden, Spain, UK: water pollution | €2.43–€4.75 billion per year | OECD |
| Global: biodiversity loss | Globally food production accounts for 60%–70% of total biodiversity loss | PBL Netherlands Environmental Assessment Agency |
| Global: climate change | $100–$250 billion per year | DARA & the Climate Vulnerable Forum and Springmann et al (2016a) |
| Global: feeding cereals to farm animals in 2050 | $1323 billion per year | Schader et al (2015); FAO, *Food waste footprint: full cost accounting* & author's calculation |
| Global: non-communicable diseases (NCD) | $735 billion per year using a cost-of-illness approach | Springmann et al (2016a) |
| EU: antimicrobial resistance | €1.5 billion per year | European Commission |
| US: antimicrobial resistance | €55 billion per year (includes treatment costs & lost productivity) | US Centers for Disease Control and Prevention |

(*Continued*)

*Table 20.2* (Continued)

| *Negative externality* | *Estimated cost* | *Source* |
|---|---|---|
| Global: loss of economic output due to antimicrobial resistance between now and 2050 on BAU basis | $1710–$2860 billion per year | Study commissioned by UK Government |
| Denmark: health-related costs of air pollution | €2.1 billion per year | Brandt et al |
| France: total cost of air pollution | €27–€41 billion per year (on assumption that, as in Denmark, 43% of air pollution costs are attributable to agriculture) | French Senate |
| EU: exposure to pesticides | $127 billion per year | TEEB for Agriculture & Food |
| US: foodborne disease | $15.5–$152 billion per year | Scharff, USDA |
| EU: campylobacter | €2.4 billion per year | EFSA |
| EU: salmonella | €3 billion per year | EFSA |
| EU: animal welfare costs of intensive pig sector | €19 billion per year | Van Drunen et al. & author's calculation |

## The need to internalise the negative externalities of livestock production

Our economic system is generally poorly equipped to take into account the impact of agriculture on factors that are not owned by anyone and for which there is no, or only a partial, market. These factors include, for example, clean air, animal welfare, climate stability, good dietary health and the need to leave sufficient and good quality water, soil and biodiversity for future generations. Such factors do not have to be paid for by farmers and consumers of food and so, in the absence of some form of intervention, are vulnerable to receiving insufficient attention.

An economic system that arbitrarily takes account of some of the costs of producing food while ignoring others is inefficient and produces undesirable outcomes such as poor levels of dietary health, erosion of agriculture's core factors of production (soil, water, biodiversity) and low standards of animal welfare. This capricious failure to take certain costs into account has produced a food system that makes unhealthy, environmentally damaging food cheaper than food that is nutritious and respects the environment and animal welfare.

The consequence of unhealthy food being cheaper in the West than healthy food is that poorer members of society find themselves having to rely on poor quality food. For example, the Faculty of Public Health states that "in the UK, the poorer people are, the worse their diet, and the more diet-related diseases they suffer from" (Faculty of Public Health, no date). Olivier De

Schutter, former United Nations (UN) Special Rapporteur on the right to food, stresses that "any society where a healthy diet is more expensive than an unhealthy diet is a society that must mend its price system" (De Schutter, 2011). This applies equally to a society where environmentally damaging, low animal welfare food is cheaper than food that respects natural resources and animals' well-being.

## Mending our food price system

A principal objective of internalising negative externalities is to achieve a better alignment between an individual's incentives and societal objectives.

A wide range of measures can be used to internalise both positive and negative externalities. Legislation, fiscal measures, codes of practice and standards set by food businesses can all internalise external costs. Taxes can be used to internalise external costs and/or to encourage or discourage certain production or consumption decisions.

A report by the WHO Europe region points out that taxation specialists recognise that the purpose of the tax system is not just to raise revenue but that it plays a role in supporting policy objectives such as health gains and healthcare cost savings (WHO Europe, 2015). It stresses that "consumers can be highly responsive to food prices and that taxation and subsidies are an effective means of influencing consumption of targeted foods".

### *Internalising the societal costs of unhealthy food and promoting healthy diets*

The UN advocates the use of fiscal measures to promote healthy diets. The UN *Political Declaration on Non-Communicable Diseases* (NCDs) identifies unhealthy diets as a key risk factor for NCDs (UN General Assembly, 2012). It urges Governments to advance interventions to reduce the impact of unhealthy diets on NCDs through, *inter alia* "fiscal measures". In his 2011 report on NCDs the UN Secretary-General identifies food subsidies and taxes as a cost-effective way of promoting healthy diets (UN General Assembly, 2011). Countries could, for example, place a tax on unhealthy foods and use the income generated to subsidise healthy foods.

Research shows that a tax on unhealthy foods, combined with the appropriate amount of subsidy on fruits and vegetables, could lead to significant health gains (Nnoaham et al., 2009). A Danish study concluded that taxes on "unhealthy" and subsidies for "healthy" food products can improve public nutrition (ATV, 2007). US research found that small price differences at the point of purchase can be highly effective in shifting consumer demand from high-calorie milk to healthier low-calorie alternatives (Khan et al., 2015). It reports that low-income consumers who are at higher risk for obesity are particularly responsive.

The WHO report referred to earlier points out that without government intervention the prices of fruit and vegetables at point of purchase are likely to exceed the socially optimal price, and the quantity sold will be below the level needed for the maximum benefit to society (WHO Europe, 2015). The report emphasises the effectiveness of subsidies in increasing the purchase of healthy foods and of taxes in decreasing the purchase of unhealthy foods. It states that the potential for positive effects might be amplified if a targeted food tax were combined with a subsidy on fruit and vegetables or other healthy foods with the subsidy being funded by the revenue raised by the tax.

Brazil has made commitments on ending obesity that include "fiscal measures (subsidies, tax reductions etc.) in order to reduce the price of healthy foods, such as fruits and vegetables" (Barros, 2017).

### *Preventing regressivity*

Taxes on food must be designed so as to avoid having an unfair impact on poorer people as a tax-related price increase will place a greater burden on them than on wealthier consumers. This can be mitigated by subsidies on healthy food so that the overall price of food does not increase.

The WHO points out that for poor socio-economic groups a food tax may lead to dietary shifts and so to improved dietary health provided that untaxed, healthy alternatives are available; such health gains may contribute to reducing health inequalities (WHO Europe, 2015). The OECD has concluded that, of all actions to prevent obesity "fiscal measures are the only intervention producing consistently larger health gains in the less well-off" across the countries studied (Sassi, 2010).

### *Impact of tax or charge can go beyond its monetary value*

The WHO points out that taxation may result in consumers becoming more aware of the unhealthy properties of certain products because of the price increase, thereby amplifying the effect of the price increase and enhancing the market for healthy products (WHO Europe, 2015).

### *Internalising the societal costs of farm use of antimicrobials*

The O'Neill report examined the case for placing a tax on the use of antimicrobials in the livestock sector (O'Neill, 2015). It advised that this would ensure that farm use would take into account the societal cost of antimicrobial use and increase the economic incentive for farmers to use alternatives, such as improved husbandry and vaccination. It said that the tax should be set at a level that discourages growth promotion (which is still used in many countries), and unnecessary prophylactic use, but that does not stop farmers from adequately treating their sick animals.

### *Internalising negative environmental externalities and promoting sustainable agriculture*

Environmental taxes are in operation in certain countries, for example, carbon/ energy taxes, sulphur taxes, leaded and unleaded petrol tax differentials, landfill taxes, pesticide taxes and fertiliser taxes. Such measures are designed to internalise the external costs of certain activities.

A UN Development Programme (UNDP) paper examines how taxes on pesticides and fertilisers can correct certain market failures (e.g. the failure to incorporate in the price of the pesticide/fertiliser its social and environmental costs) and can forestall increases in the use of the most harmful pesticides and fertilisers (UNDP, 2017). Such taxes can lead to savings in health budgets (including lost productivity) and reduced expenditure in restoration of degraded land and natural resources.

The UNDP paper states that from an economic perspective, a differentiated tax that takes account of the damage to the environment and human well-being caused by different types of pesticides/fertilisers is the preferred solution since it provides more targeted price signals to the market and more adequately reflects marginal damages.

The paper points out that the revenue generated by such taxes could be earmarked to mitigating the environmental impacts of pesticides and fertilisers, adopting more sustainable agriculture practices and otherwise contributing to the achievement of a country's sustainable development goals. It stresses that these taxes are "more appropriate where the objective is to facilitate a smooth transition to more sustainable practices through market mechanisms".

Such taxes should be seen not as a substitute for legislation but as complementing regulations. The UNDP paper states, "an example is seen in France where a combined system is in place in which a reduced tax rate is imposed on pesticides that are allowed in organic farming, while the regular tax rate is imposed on other pesticides, and a total ban is imposed on some widely used pesticides that are considered to harm bees" (UNDP, 2017).

### *Joint health and environmental benefits of taxing certain foods*

Many studies show that a dietary pattern higher in plant-based foods and lower in animal-based foods would be beneficial for the environment and public health and would reduce GHG emissions (Aston et al., 2012; Springmann et al., 2016a; Westhoek et al., 2015).

Springmann et al. (2016b) show that levying GHG taxes on food commodities could, if appropriately designed, both lower GHG emissions and promote health in high-income countries, as well as in most low- and middle-income countries (Springmann et al., 2016b). However, taxes on food must spare food groups that are beneficial for health from taxation and use the tax revenues for health promotion and subsidising the consumption of fruit and vegetables.

The UN Standing Committee on Nutrition states that policies to make diets healthier and sustainable with low environmental impacts include economic incentives (UNSCN, 2017). They say this could involve taxing unhealthy food and subsidising or providing economic incentives for the consumption of healthier food.

### *Internalising the societal costs of the production and consumption of animal products*

Similar approaches could be taken in the field of livestock production. One approach to internalising the externalities of meat production – that is, including them in the price of meat – is the introduction of a Pigouvian Tax that reflects the cost of the negative externalities.[1] Such a tax would correct the market failure due to externalities. The Dutch study referred to earlier states that the average rate of the Pigouvian Tax should be at least €2.06 for 1 kg of conventionally produced pork, which is 31% of the consumer price in the Netherlands at the time of the study.

A Swedish study considers three meat products, cattle, chicken and pork, and three pollutants generating environmental damages, GHG, nitrogen and phosphorus (Säll and Gren, 2012). The study examines taxes on meat products corresponding to the environmental damage caused by the different products; these amount to 28%, 26% and 40% of the price per kg of beef, pork and poultry, respectively, in 2009. The study calculates that a simultaneous introduction of taxes on all three meat products can decrease emissions of GHGs, nitrogen, phosphorus and ammonia by at least 27%.

Taxes on meat should not apply to all meat but only to that which is produced industrially as it is this meat that is responsible for most of the sector's adverse environmental impacts and most of its use of antimicrobials and that generally is of lower nutritional quality than free-range meat. Moreover, the industrial livestock sector has inherent severe deficiencies for animal welfare. In contrast, extensive indoor systems and outdoor rearing have the potential, if well-designed and well-managed, to deliver good welfare outcomes. Accordingly, taxes should not be placed on meat from well-managed pasture-based herds, integrated rotational crop-livestock systems or extensive indoor or free-range systems.

Revenue raised from such taxes should be used to subsidise healthy foods such as fruits and vegetables, legumes and whole grains as it is crucial from the viewpoint of social equity that the overall price of food does not increase.

## Using fiscal measures positively

Tax measures should be used not just to reflect the cost of negative externalities but the revenue raised should be used to lower the costs of particular farming practices and certain foods. They should be used to make healthy food produced to high environmental and animal welfare standards economically attractive for both farmers and consumers.

### *Supporting farmers*

Farmers producing to high environmental and animal welfare standards could be compensated for the extra costs involved by subsidies and tax breaks. When calculating net profits for tax purposes, more generous capital allowances could be given to investments in high-quality farming. Governments already uses differential capital allowances to reward activities that they wish to encourage; for example, enhanced capital allowances are given in some countries for businesses that use environmentally beneficial technologies. Moreover, an extra tranche of farmers' taxable income could be tax-free when they employ specified animal welfare or environmental practices. These tax breaks could be paid for by the revenue raised from placing taxes on the inputs of industrial agriculture such as chemical fertilisers and pesticides.

### *Supporting consumers*

Taxes should be placed on unhealthy, inhumanely produced food with the revenue raised being used to subsidise the price of healthy food produced to high environmental and animal welfare standards. In countries which charge value-added tax (VAT) on food, the price paid by consumers for such food could be reduced by placing a lower or nil VAT rate on such food.

Studies show varying results as to how effective fiscal measures can be in influencing consumer behaviour. However, a report by Chatham House and the Food Climate Research Network (FCRN) stresses that "lack of evidence should not be used as an excuse for policy inaction. Indeed policy inaction leads to a paucity of empirical evidence. Trials and experimentation particularly based on some of the more politically challenging fiscal and regulatory approaches discussed are essential. As noted, robust monitoring and evaluation processes need to be in place so that impacts in the short, medium and longer term can be understood. In this way the evidence base is built and policies progressively refined and improved" (Garnett et al., 2015).

Fiscal measures cannot on their own reshape our food system into one that delivers high-quality food. They must be implemented in conjunction with other strategies and policies that aim to improve our food system including regulation, voluntary initiatives by food businesses, supportive public procurement and consumer information.

The report by Chatham House and FCRN stresses that while they have important roles to play, the restructuring of our food system cannot be left to "industry goodwill or enlightened self interest" (Garnett et al., 2015). The report highlights the need for governments' non-interventionist approach to be replaced by a willingness to set a strong policy, regulatory and fiscal framework. It emphasises that governments must govern and must be prepared to step in and lead.

## Annex: how the costs in the final column of Table 20.1 were calculated

The first two columns are drawn from Figure 1 in Schader et al. (2015). The third column subtracts the figures in the second column from those in the first column.

The final column calculates the costs arising from the detrimental impacts that are quantified in the third column. These calculations have been made using the data in Table 20.2 and elsewhere in the FAO report *Food waste footprint: full cost accounting* (FAO, 2014b).

*GHG emissions*: Table 20.2 of the FAO report suggests a figure of $113/tonne CO2e. 2.4 Gigatonnes CO2e of emissions entails a cost of $271 billion per year.

*Freshwater use (for irrigation)*: Table 20.2 of the FAO report suggests a figure of $0.1/m$^3$ for water use and $1.15/m$^3$ in respect of water scarcity making a total of $1.25/m$^3$ (see pages 41–42 of the FAO report). 460 km$^3$ of freshwater use entails an annual cost of $575 billion.

*Nitrogen-surplus*: Table 20.2 of the FAO report suggests a figure of $0.286/kg N leached in respect of eutrophication impact on water and $0.0245/kg N applied in respect of eutrophication impact on biodiversity. The annual N surplus of 56.6 million tonnes entails a cost of $17.5 billion per year.

*Phosphorus-surplus*: Table 20.2 of the FAO report suggests a figure of $12.32/kg P leached in respect of eutrophication impact on water and $0.26/kg P applied in respect of eutrophication impact on biodiversity. The annual P surplus of 25.6 million tonnes entails a cost of $322 billion per year.

*Pesticide use*: The paper by Schader et al. does not specify a unit for measuring pesticide use. However it shows that pesticide use would be much higher in 2050 under BAU than if the use of cereals as feed was ended. The FAO paper on food waste estimates the adverse health effects due to pesticide exposure attributable to the production of food that is then wasted to be $153 billion per year.

The FAO paper estimates that approximately one-third of food produced for human consumption is lost or wasted. Studies show that for every 100 calories of cereals fed to animals just 17–30 (an average of 23.5) calories enter the human food chain as meat (Lundgvist et al., 2008; Nellemann et al., 2009). The effect of this is that 76.5% of cereals fed to animals is lost due to animals' poor conversion of cereals into meat. 36% of global cereal production is fed to animals; 76.5% of this is lost. This means that 27% of global cereal production is lost by being used as animal feed.

To sum up, 33% of food is lost or wasted in the conventional sense (e.g. post-harvest losses, food discarded by consumers). A roughly similar figure – 27% – of global cereal production is lost by being used as animal feed. The FAO estimates the adverse health effects due to pesticide exposure attributable to the production of food that is then wasted to be $153 billion per year. A substantial proportion of pesticides are used in cereal production.

This paper takes a cautious approach and presumes that if the pesticide impact on health of growing food that is then wasted is $153 billion per year, the pesticide impact on health of that proportion of cereals used as animal feed that does not produce meat is $38 billion per year (25% of $153 billion).

*Deforestation*: Table 20.2 of the FAO report suggests a figure of $1611 per hectare of forest lost for loss of ecosystem services from deforestation. The annual loss of 700,000 hectares of forest entails a cost of $1.1 billion per year.

*Soil erosion from water*: Table 20.2 of the FAO report suggests a figure of $21.54/ton of soil lost by water erosion. The annual loss of 4.6 billion tonnes soil entails a cost of $99 billion per year.

## Note

1 Wikipedia describes a Pigouvian tax as a tax levied on a market activity that generates negative externalities. The tax is intended to correct the market outcome. In the presence of negative externalities, the social cost of a market activity is not covered by the private cost of the activity. In such a case, the market outcome is not efficient and may lead to overconsumption of the product. A Pigouvian tax equal to the negative externality is thought to correct the market outcome back to efficiency.

## References

Aichi, L., 2015. Pollution de l'air: le coût de l'inaction. Rapport fait au nom de la commission d'enquête (1) sur le coût économique et financier de la pollution de l'air. La CE Coût Économique et Financier de la Pollution de L'air. 610 www.senat.fr/rap/r14-610-1/r14-610-1.html. Accessed 11 April 2018.

Aston, L.M., Smith, J.N. and Powles, J.W., 2012. Impact of a reduced red and processed meat dietary pattern on disease risks and greenhouse gas emissions in the UK: a modelling study. *BMJ Open. 2*(5), 2e001072. http://bmjopen.bmj.com/content/2/5/e001072.full.pdf+html. Accessed 11 April 2018.

ATV (Danish Academy of Technical Sciences), 2007. *Economic nutrition policy tools—useful in the challenge to combat obesity and poor nutrition?*

Bailey, R., Froggatt, A. and Wellesley, L., 2014. *Livestock – climate change's forgotten sector: global public opinion on meat and dairy consumption.* Chatham House. www.chathamhouse.org/publication/livestock-climate-change-forgotten-sector-global-public-opinion-meat-and-dairy. Accessed 10 April 2018.

Bajželj, B., Richards, K.S., Allwood, J.M., Smith, P., Dennis, J.S., Curmi, E. and Gilligan, C.A., 2014. Importance of food-demand management for climate mitigation. *Nature Climate Change. 4*, pp. 924–929. www.nature.com/doifinder/10.1038/nclimate2353. Accessed 10 April 2018.

Barros, R., 2017. Letter from Minister of Health, Federative Republic of Brazil to WHO. www.unscn.org/en/topics/un-decade-of-action-on-nutrition?idnews=1684. Accessed 11 April 2018.

Batz, M.B., Hoffmann S. and Morris Jr, J.G., 2011. *Ranking the risk: the 10 pathogen-food combinations with the greatest burden on public health.* Emerging Pathogens Institute, University of Florida. https://folio.iupui.edu/bitstream/handle/10244/1022/72267report.pdf?sequence=1. Accessed 11 April 2018.

Brandt, J., Silver, J.D., Christensen, J.H., Andersen, M.S., Bønløkke, J.H., Sigsgaard, T., Geels, C., Gross, A., Hansen, A.B., Hansen, K.M., Hedegaard, G.B., Kaas, E. and Frohn, L.M., 2011. *Assessment of health-cost externalities of air pollution at the national level using the EVA model system.* Centre for Energy, Environment and Health Report Series www.researchgate.net/publication/284676933_Assessment_of_Health-Cost_Externalities_of_Air_Pollution_at_the_National_Level_using_the_EVA_Model_System. Accessed 11 April 2018.

Cassidy, E.S., West, P.C., Gerber, J.S. and Foley, J.A., 2013. Redefining agricultural yields: from tonnes to people nourished per hectare. University of Minnesota. *Environmental Research Letters, 8*(3), 034015. www.iopscience.iop.org/article/10.1088/1748-9326/8/3/034015/pdf. Accessed 10 April 2018.

CDC, 2013. Antibiotic resistance threats in the United States. U.S. Department of Health and Human Services. www.cdc.gov/drugresistance/pdf/ar-threats-2013-508.pdf. Accessed 11 April 2018.

De Schutter, O., 2011. *Report of the special rapporteur on the right to food.* Human Rights Council, UN. A/HRC/19/59 www.ohchr.org/Documents/HRBodies/HRCouncil/RegularSession/Session19/A-HRC-19-59_en.pdf. Accessed 11 April 2018.

Dodds, W.K., Bouska, W.W., Eitzmann, J.L., Pilger, T.J., Pitts, K.L., Riley, A.J., Schloesser, J.T. and Thornbrugh, D.J., 2009. Eutrophication of U.S. freshwaters: analysis of potential economic damages. *Environmental Science & Technology, 43*(1), pp. 12–19. www.pubs.acs.org/doi/full/10.1021/es801217q. Accessed 10 April 2018.

Edmondson, J.L., Davies, Z.G., Gaston, K.J. and Leake, J.R., 2014. Urban cultivation in allotments maintains soil qualities adversely affected by conventional agriculture. *Journal of Applied Ecology, 51*, pp. 880–889. doi:10.1111/1365-2664.12254. Accessed 11 April 2018.

EFSA BIOHAZ Panel (EFSA Panel on Biological Hazards), 2011. Scientific opinion on campylobacter in broiler meat production: control options and performance objectives and/or targets at different stages of the food chain. *EFSA Journal. 9*(4), p. 141. www.efsa.europa.eu/en/efsajournal/pub/2105. Accessed 11 April 2018.

EFSA, undated. *Salmonella.* www.efsa.europa.eu/en/topics/topic/salmonella.htm. Accessed 11 April 2018.

European Commission, 2011a. *Commission staff working paper. Analysis associated with the roadmap to a resource efficient Europe part II, SEC 1067 final.* www.ec.europa.eu/environment/resource_efficiency/pdf/working_paper_part2.pdf. Accessed 10 April 2018.

European Commission, 2011b. *Action plan against the rising threats from antimicrobial resistance.* https://ec.europa.eu/health/amr/sites/amr/files/ev_20151022_co32_en.pdf. Accessed 10 April 2018.

European Commission, 2012. *Consultation paper: options for resource efficiency indicators.* http://ec.europa.eu/environment/consultations/pdf/consultation_resource.pdf. Accessed 10 April 2018.

European Environment Agency, 2015. European briefings: biodiversity. www.eea.europa.eu/soer-2015/europe/biodiversity. Accessed 10 April 2018.

Faculty of Public Health, undated. *Briefing statement: food poverty and health.* www.fph.org.uk/uploads/bs_food_poverty.pdf. Accessed 11 April 2018.

FAO, 2014a. *Global plans of action endorsed to halt the escalating degradation of soils.* www.fao.org/news/story/en/item/239341/icode/. Accessed 10 April 2018.

FAO, 2014b. *Food wastage footprint: full-cost accounting.* www.fao.org/3/a-i3991e.pdf. Accessed 10 April 2018.

FAO, 2015. *Natural capital impacts in agriculture.* www.fao.org/fileadmin/templates/nr/sustainability_pathways/docs/Natural_Capital_Impacts_in_Agriculture_final.pdf. Accessed 10 April 2018.

Foresight, 2011. *The future of food and farming: final project report.* The Government Office for Science, London. www.gov.uk/government/publications/future-of-food-and-farming. Accessed 10 April 2018.

Garnett, T., Mathewson, S., Angelides, P. and Borthwick, F., 2015. *Policies and actions to shift eating patterns: what works?* Chatham House and Food Climate Research Network. https://fcrn.org.uk/sites/default/files/fcrn_chatham_house_0.pdf. Accessed 11 April 2018.

Gerber, P.J., Steinfeld, H., Henderson, B., Mottet, A., Opio, C., Dijkman, J., Falcucci, A. and Tempio, G., 2013. *Tackling climate change through livestock—A global assessment of emissions and mitigation opportunities.* FAO. www.fao.org/3/a-i3437e.pdf. Accessed 10 April 2018.

Hoffmann, S., Maculloch, B. and Batz, M., 2015. *Economic burden of major foodborne illnesses acquired in the United States.* USDA Economic Research Service. Economics Information Bulletin No. 140. www.ers.usda.gov/webdocs/publications/43984/52807_eib140.pdf. Accessed 11 April 2018.

Khan, R., Misra, K. and Singh, V., 2015. Will a fat tax work? *Marketing Science.* pp. 10–26. https://pubsonline.informs.org/doi/10.1287/mksc.2015.0917. Accessed 11 April 2018.

Kluser, S., Neumann, P., Chauzat, M. and Pettis, J.S., 2010. *Global honey bee colony disorders and other threats to insect pollinators.* United Nations Environment Programme. https://tinyurl.com/UNEP-global-bee-colony. Accessed 10 April 2018.

Kok, M., Alkemade, R., Bakkenes, M., Boelee, E., Christensen, V., van Eerdt, M., van der Esch, S., Janse, J., Karlsson-Vinkhuyzen, S., Kram, T., Lazarova, T., Linderhof, V., Lucas, P., Mandryk, M., Meijer, J., van Oorschot, M., Teh, L., van Hoof, L., Westhoek, H. and Zagt, R., (Eds), 2014. How sectors can contribute to sustainable use and conservation of biodiversity. *PBL Netherlands Environmental Assessment Agency.* 79, p. 230. www.pbl.nl/sites/default/files/cms/publicaties/PBL_GBO4_Sectoral%20mainstreaming_low_res.pdf. Accessed 10 April 2018.

Lundqvist, J., de Fraiture, C. and Molden, D., 2008. *Saving water: from field to fork – curbing losses and wastage in the food chain. SIWI policy brief.* SIWI. www.siwi.org/publications/saving-water-from-field-to-fork-curbing-losses-and-wastage-in-the-food-chain/. Accessed 10 April 2018.

Lyne, A., Jørgensen, F., Little, C., Gillespie, I., Owen, R., Newton, J. and Humphrey, T., 2007. *Project B15019: review of current information on campylobacter in poultry other than chicken and how this may contribute to human cases.*

Mccubbin, D.R., Apelberg, B.J., Roe, S. and Divita, F., 2002. Livestock ammonia management and particulate-related health benefits. *Environmental Science & Technology. American Chemical Society. 36*(6), pp. 1141–1146. https://pubs.acs.org/doi/abs/10.1021/es010705g. Accessed 11 April 2018.

Mekonnen, M.M. and Hoekstra, A.Y., 2012. A global assessment of the water footprint of farm animal products. *Ecosystems. 15*(3), pp. 401–415. https://link.springer.com/article/10.1007/s10021-011-9517-8. Accessed 10 April 2018.

Moxey, A., 2012. *Agriculture and water quality: monetary costs and benefits across OECD countries.* OECD. www.oecd.org/tad/sustainable-agriculture/49841343.pdf. Accessed 10 April 2018.

Nellemann, C., MacDevette, M., Manders, T., Eickhout, B., Svihus, B., Prins, A. G. and Kaltenborn, B. P. (Eds), 2009. *The environmental food crisis – The environment's role in averting future food crises. A UNEP rapid response assessment.* United Nations Environment Programme, GRID-Arendal. www.gwp.org/globalassets/global/toolbox/references/the-environmental-crisis.-the-environments-role-in-averting-future-food-crises-unep-2009.pdf. Accessed 10 April 2018.

Nnoaham, K.E., Sacks, G., Rayner, M., Mytton, O. and Gray, A., 2009. Modelling income group differences in the health and economic impacts of targeted food taxes and subsidies. *International Journal of Epidemiology. 38*(5) pp. 1324–1333.

O'Neill, J., 2014. *Antimicrobial resistance: tackling a crisis for the health and wealth of nations.* The Review on Antimicrobial Resistance. https://tinyurl.com/AMR-antimicrobial-review. Accessed 11 April 2018.

O'Neill, J., 2015. Antimicrobials in agriculture and the environment: reducing unnecessary use and waste. The Review on Antimicrobial Resistance. https://tinyurl.com/environ-antimicrob-resistance. Accessed 11 April 2018.

Pretty, J.N., Brett C., Gee D., Hine R.E., Mason C.F., Morison J.I.L., Rayment M.D., van der Bijl G. and Dobbs T., 2001. Policy challenges and priorities for internalizing the externalities of modern agriculture. *Journal of Environmental Planning and Management, 44*(2), pp. 263–283.

Säll, S. and Gren, I.-M., 2012. *Green consumption taxes on meat in Sweden.* Working paper 10/2012. Swedish University of Agricultural Sciences. https://pub.epsilon.slu.se/9294/1/sall_s_121214.pdf. Accessed 11 April 2018.

Sutton, M.A., Howard, C.M., Erisman, J.W., Billen, G., Bleeker, A., Grennfelt, P., van Grinsven, H. and Grizzetti, B. (Eds), 2011. *The European Nitrogen Assessment.* Cambridge University Press, New York.

Sassi, F., 2010. *Obesity and the economics of prevention: fit not fat.* OECD. http://s3.amazonaws.com/zanran_storage/www.eaca.be/ContentPages/956222899.pdf. Accessed 11 April 2018.

Schader, C., Muller, A., El-Hage Scialabba, N., Hecht, J., Isensee, A., Erb, K., Smith, P., Makkar, H.P., Klocke, P., Leiber, F., Schwegler, P., Stolze, M. and Niggli, U., 2015. Impacts of feeding less food-competing feedstuffs to livestock on global food system sustainability. *Journal of the Royal Society Interface. 12*(20150891) http://dx.doi.org/10.1098/rsif.2015.0891. Accessed 10 April 2018.

Scharff, R.L., 2010. *Health-related costs from foodborne illness in the United States.* The Produce Safety Project, Georgetown University. www.producesafetyproject.org. Accessed 11 April 2018.

Steinfeld, H., Gerber, P., Wassenaar, T., Castel, V., Rosales, M. and de Haan, C., 2006. *Livestock's long shadow: environmental issues and options.* UN Food and Agriculture Organisation. www.europarl.europa.eu/climatechange/doc/FAO%20report%20executive%20summary.pdf. Accessed 10 April 2018.

Springmann, M.H., Godfray, C.J., Rayner, M., Scarborough, P. and Tilman, D. (Ed), 2016a. Analysis and valuation of the health and climate change cobenefits of dietary change. PNAS. *113*(15) pp. 4146–4151. www.pnas.org/content/113/15/4146. Accessed 10 April 2018.

Springmann, M.H., Mason-D'Croz, D., Robinson, S., Wiebe, K., Godfray, C.J., Rayner, M., Scarborough, P., 2016b. Mitigation potential and global health impacts from emissions pricing of food commodities. *Nature Climate Change.* 7, pp. 69–74. www.nature.com/articles/nclimate3155. Accessed 11 April 2018.

TEEB, 2015. *TEEB for agriculture & food interim report.* United Nations Environment Programme, Geneva, Switzerland. http://img.teebweb.org/wp-content/uploads/2016/01/TEEBAgFood_Interim_Report_2015_web.pdf?utm_source=

website&utm_medium=report&utm_campaign=TeebAgriFoodInterimReport. Accessed 11 April 2018.

Trasande, L., Zoeller, R., Hass, U., Kortenkamp, A., Grandjean, P., Myers, J., DiGangi, J., Bellanger, M., Hauser, R., Legler, J., Skakkebaek, N. and Heindel, J., 2015. Estimating burden and disease costs of exposure to endocrine-disrupting chemicals in the European Union. *The Journal of Clinical Endocrinology and Metabolism. 100*(4), pp. 1245–1255. www.ncbi.nlm.nih.gov/pmc/articles/PMC4399291/. Accessed 11 April 2018.

Turbé, A., De Toni, A., Benito, P., Lavelle, P., Lavelle, P., Ruiz, N., Van der Putten, W.H., Labouze, E. and Mudgal, S., 2010. *Soil biodiversity: functions, threats and tools for policy makers*. European Commission. http://ec.europa.eu/environment/archives/soil/pdf/biodiversity_report.pdf. Accessed 10 April 2018.

UNDP, 2017. *Taxes on pesticides and chemical fertilizers*. www.undp.org/content/sdfinance/en/home/solutions/taxes-pesticides-chemicalfertilizers.html. Accessed 11 April 2018.

UN General Assembly, 2012. *Political declaration of the high-level meeting of the general assembly on the prevention and control of non-communicable diseases*. A/RES/66/2. www.who.int/nmh/events/un_ncd_summit2011/political_declaration_en.pdf. Accessed 11 April 2018.

UN General Assembly, 2011. *Report of the UN Secretary-General: prevention and control of non-communicable diseases*. A/66/83. www.who.int/nmh/publications/2011-report-of-SG-to-UNGA.pdf. Accessed 11 April 2018.

UNSCN, 2017. *Sustainable diets for healthy people and a healthy planet*. www.unscn.org/uploads/web/news/document/Climate-Nutrition-Paper-EN-WEB.pdf. Accessed 11 April 2018.

Van Drunen, M., van Beukering, P. and Aiking, H., 2010. *The true price of meat. Report W10/02aEN*. Institute for Environmental Studies, VU University, Amsterdam, The Netherlands. file:///C:/Users/volunteer/Downloads/The_true_price_of_meat.pdf. Accessed 11 April 2018.

Wentworth, J., 2013. *Reversing insect pollinator decline. Number 442*. The Parliamentary Office of Science and Technology. www.parliament.uk/business/publications/research/briefing-papers/POST-PN-442/reversing-insect-pollinator-decline. Accessed 10 April 2018.

Westhoek, H., Lesschen, J.P., Leip, A., Rood, T., Wagner, S., De Marco, A., Murphy-Bokern, D., Pallière, C., Howard, C.M., Oenema, O. and Sutton, M.A., 2015. *Nitrogen on the table: the influence of food choices on nitrogen emissions and the European environment* (European Nitrogen Assessment Special Report on Nitrogen and Food.) Centre for Ecology & Hydrology, Edinburgh, UK.

WHO Europe, 2015. *Using price policies to promote healthier diets*. www.euro.who.int/__data/assets/pdf_file/0008/273662/Using-price-policies-to-promote-healthier-diets.pdf. Accessed 11 April 2018.

World Cancer Research Fund/American Institute for Cancer Research, 2011. *Continuous update project interim report summary*. Food, Nutrition, Physical Activity and the Prevention of Colorectal Cancer.

World Health Organisation, 2012. *The evolving threat of antimicrobial resistance: options for action*. www.who.int/iris/handle/10665/44812. Accessed 10 April 2018.

# 21 Changing the world, one meal at a time

*Martin Palmer*

At the Alliance of Religions and Conservation (ARC) we work with every major religion around the world – Baha'i, Buddhists, Christians, Daoists, Hindus, Jains, Jews, Muslims, Sikhs, Shinto, Zoroastrians – helping them develop environmental programmes based on their teachings and traditions. But we haven't the slightest interest in what they have in common. We are not an interfaith organisation. People belong to a specific tradition within a faith because of what it is that is distinctive and special about it, not because it is the same as everyone else. We are interested in why the followers of Lord Krishna in northern India will protect their local river and why, with different stories, the followers of Lord Jagannath in eastern India will protect their local forests. Each is rooted in their own specific stories, landscape and history. Within the major faiths are many different traditions. Over the past 30 years we have worked with about 15 Buddhist traditions for example, and what is of environmental concern to Mongolian Buddhists is somewhat different from what troubles Sri Lankan Buddhists.

Essentially what we work with are belief systems. Belief systems with profound principles of values. And it is those values which then inform the choices and decisions these faith traditions make and embrace. They are, in many cases, the fruits of centuries (or even millennia) of thought, philosophy, experience, poetry, storytelling and insight. What we call the Wisdom of the Faiths developed over centuries of dealing with human nature as well as what some people might describe as "above nature" or, literally, "supernatural".

But there is another aspect of the faiths. The business of faiths. Religions are huge business. They run, founded or are actively connected with, over 50% of all schools worldwide and according to United Nations International Children's Emergency Fund (UNICEF) that figure is 64% in sub-Saharan Africa. They own about 8% of the habitable surface of the planet – and 15% of the planet is considered "sacred". There are sacred mountains, sacred rivers, sacred forests. And there are sacred cities. These places are often not owned by the faiths but they are profoundly influenced and protected by being considered sacred.

Religions (and their pension and other funds) together make up the fourth largest investing group in the world. In October 2017 we brought together representatives of some of the major religious investment groups in the town of Zug in Switzerland. Eight religions were represented, from Buddhism to Shintoism and discussed how that investment can be invested in environmental and sustainable development. I say "environmental and sustainable" development because our experience is that sustainable development is rarely also environmental.

Usually the environment bit is planting a tree at the conclusion of the development. That is not what we mean by environmental. The Zug event marks a massive shift. Faiths have known for centuries what they are against. Islam against the charging of interest; Quakers against armaments; Jains against the taking of any form of life; Evangelicals against alcohol and gambling... This time we have asked them a simple question. "We know what you are against. What are you for?" What will the faiths become pro-active investors in to make a better world? That question is beginning to turn the world upside down.

The faiths already contribute to the well-being of the planet, not least in relation to food. They feed hundreds of millions of people every day. The Sikhs, for example, feed some 30 million people a day in India and around the world through the free vegetarian kitchens – langar – in their gurdwaras, places of worship. Faiths feed people in hospitals, schools, refugee centres, through relief work and through celebrating the great festival days or people's rites of passage. They purchase vast quantities of food every day and they also grow a great deal of it (e.g. the Church of England owns about 11% of the farmland in England). Different faiths also have important teachings about food, including –which is, of course, at the core of Compassion in World Farming (Compassion) – the relationship of our values with the animals we eat or use for our farming.

I have had the joy of working with Compassion for almost 30 years, and I love the fact that, unlike almost any other environmental organisation, it retains in its title a relationship word, an emotional word: compassion. Because that is what motivates us. Facts and figures are, of course, useful but without a sense of relationship, without a greater story, they are largely meaningless. Long may Compassion, and we as its friends, be the "compassionate movement".

In the book of Lieh Zi (probably compiled in the 4th century and one of the great classics of Daoism from China), there is this story:

> A wealthy merchant was going on a long journey and before he left he hosted a banquet. No expense was spared. There were fish dishes, chicken and duck, beef and pork, rare animals were served as well. It was a great feast and people enjoyed themselves. At the end the merchant stood up and said "Isn't it wonderful the way Nature has created all these delicious foods for us and keeps them fresh in the fish, birds and animals until we want to eat them!" All the guests cheered. Except for one ten-year-old boy who stood up and said "No! You are wrong!" The merchant and his guests were astonished and the merchant asked the boy what he meant.

> "Look," said the boy. "If they get half a chance, wolves and tigers eat our flesh and crunch our bones. And mosquitoes bite us every chance they get, and drink our blood. Following the logic of your thinking, this would mean we were created to feed wolves, tigers and mosquitoes. "All life is created for its own good, not just to be of use to us," the boy said. And the sages wrote it down.

Now when that story was told about 1,600 years ago, it was told partly as a joke. By pointing out the absurdity of a human-centric view of the world, and the folly of seeing ourselves as apart from the rest of nature rather than as just part of nature, it was probably mildly mocking of Confucianism. But it was also profoundly truthful. And today more than ever many of us tend to live inside that story of being apart from nature and tell ourselves that everything in this wonderful awesome, compassionate universe was created just for us.

A phrase I especially hate – because it believes this dangerous myth while pretending to care for the natural world – is "eco-system deliverables". This basically suggests the purpose of the ecological systems of the world (created over billions of years, and with their untold numbers of forms of life and planets, stars and wonders) has been for us. So we can drive our cars, eat what we want, burn fossil fuels and heat our homes, offices and conference centres. This is the reductionist world view that we are now living within and its values are not only valueless. They are killing us.

As someone whose professional world includes not just the environmental movement but also the major faiths of the world, I have become used to people telling me they have the Truth. Telling me that if I do as they say, believe what they believe and follow their teachings, not only I but the whole world will be saved. They will brook no argument against their certainty and any who, for whatever reason, disagree are condemned as heretics, non-believers. What we commonly call fundamentalism. And do you know? Some of the faiths are almost as bad. Because what I am talking about is the various sects within the environmental movement.

The core problem is that each sect – or non-governmental organisation (NGO), if you want to use technical language – tends to believe it has the one true story or interpretation of the facts. If only we would all agree on this story and this interpretation of the facts, the world would be saved, I have been told on many occasions by staff from different environment organisations.

The trouble is that belief in one story, one set of truths that will save the world, has proved pretty disastrous in the past because it leaves no space for other ways of thinking or being. Whether that one story was Christianity, Islam or Buddhism. Or capitalism or Marxism, or democracy or scientific socialism… the trouble is that unless people realise what is happening, it leads to describing any other point of view as Wrong.

I think it's time for many stories. Actually the world's narrative has always been pluralist despite the best (and worst) efforts of those who wanted it to be monolithic. If God in evolution delights in diversity – so essential for life to grow, develop and change – then I am pretty sure God loves diversity in beliefs, stories and world views for exactly the same reason. We need to learn and to delight in playing with different models of reality. We need to see diversity not as a threat of division but as a source of inspiration and excitement. And that includes different models of who and what we are today and want to be in the future.

In 2015 ARC was asked to co-host with the United Nations (UN) the main meeting between the faiths and the UN about the launch of the Sustainable Development Goals (SDGs) at the end of that year, giving a vision of what the world could be like in the future – a world without discrimination or poverty, with clean air, water, education and tolerance. To the surprise of the UN, many of the faith representatives pointed out that the SDGs, clothed in good values that they are, only envisage one economic model – basically consumerist capitalism. They suggested that the aim of the SDGs could be essentially to make consumerist capitalism a little bit nicer. And they pointed out that religious organisations had, in a way, helped to run the planet for the past few thousand years and (with some monumental ups and downs) had mostly done so with very different economic models. They asked for pluralism in economic thinking. And for generosity, love and of course compassion, to be included.

At the heart of this view, and coming especially from faiths such as Daoism, Hinduism and Shintoism, is the belief that we are a part of, not apart from, the rest of Nature. You could also call it evolution, Creation, there are many words. There is a sense from those faiths that we need to dethrone ourselves from being the centre. We need other ways of seeing both ourselves and the other elements of life itself.

One way is to look at other creation stories. The Abrahamic-Scientific version suggests that humans are somewhere near the top of the evolutionary tree. The ancient Chinese have a story that can put it in perspective. The Chinese have a creation story that gives us human beings a much lesser status.

In an ancient Chinese version of the Big Bang, life begins suddenly when the Void is split by a flash of light. Yin and yang are created – light and dark, male and female, heaven and earth. At that moment a being called Pan Ku is created. When he is born he is the size of a human being, but he grows every day for 18,000 years until he fills the space between Heaven and Earth. When he dies all of life comes from his body. His flesh becomes the soil; his bones the mountains. The rivers and oceans come from his blood and the hairs on his body become the trees and plants. And where do we come from. We human beings? The Chinese story says that we come from the parasites on his bottom! A considerably less exalted version than the Western one.

We urgently need to remove ourselves from the centre of the story because otherwise we will destroy ourselves and take down much of the rest of life on earth. We have to stop treating the planet as if it were a supermarket where we can take whatever we want and someone else will refill the shelves.

October 4th is the feast day of St Francis, patron saint of animals, which symbolically marks the close of a new annual liturgical cycle in the churches called Creationtide, which begins each year on September 1st, St Giles' Day.

Eight hundred years ago St Francis spoke of Sister Sparrow and Brother Wolf, and he described the earth as our Mother. We are part of a great family but we can only regain our place within that family if we dethrone ourselves – and perhaps even ask forgiveness of the rest of creation for our arrogance.

So how do we change? It has become very fashionable in environmental movements to go on and on about behaviour change. However, as a movement, environmentalism is lagging behind many of the faiths in terms of knowing how to inspire that change. It still tends to think that fear, sin, guilt and blame will make people change. Well it doesn't. The faiths know that real change is slow – Jesus taught this in his parable of the sower (Luke ch 8 vs 4–15) and the Buddha built this into his Four Noble Truths. Faiths think in generations.

This is important in their attitude of religions towards investments. Twenty-five years is normal for them – over against the usual three to five years. Real change is a slow incremental process. It is about building relationships. And it is about increasing understanding. A tiny example of this is the Jewish traditional teaching about eating chicken. First live with the chicken for three days. Then see if you want to eat it.

The environmental movement needs to learn to ask questions, not always tell people the answer. I am reminded of the famous poster outside a church which said, "Jesus Christ is the answer!" under which someone had written "Yes, but what was the question?" I feel that often we try to give an answer without listening to what people really want to ask. My advice to any secular environmentalists is learn to ask. And then listen.

As a charity we try to use faith-based hotels and guest houses, retreat centres and such like when we travel and stay places. For about a decade we have been staying at the Methodists' International Centre (MIC) near Euston whenever we have meetings in London. This was created in the early 1960s. It was a response to a problem experienced by many overseas students, especially from Africa, when they came to study in Britain. It was often hard for them to get accommodation in student digs because they were black and people were prejudiced. As a response, the Methodists built centres for hospitality, housing and welcome.

Times have changed, thank God, so these places of Christian hospitality are no longer needed in the same way. The MIC dedicated a floor to very nice hotel style rooms, as well as the remaining student rooms. One day at breakfast we asked the manager, whom we had come to know, a simple question. Are these eggs free range? "Oh no", he said. "It's too expensive and anyway no-one is particularly interested". We replied that we were interested, that we would be willing to pay a bit extra, and we would wager that most people staying there would pay that extra too. But more than that, we also asked how the Methodists could square serving factory farmed eggs, non-compassionate eggs, with their faith values? The Methodists had issued an environmental statement a couple of years before and there was surely an inconsistency between what they preached and believed, and what they were doing.

That was all it took. One question over one breakfast. Within a year the Methodists had done a full environmental audit not just of their food purchasing policy but also energy, waste and other uses of resources. They undertook a complete transformation of their entire purchasing policy; launched a massive rebuild using only renewal energy and things like FSC wood and brought in free range, local and organic principles for all their food purchasing. And composting and recycling too. Two years later their newly renamed hotel "The Wesley" was declared Britain's first fully Ethical Hotel. Now they are rolling out a chain of Methodist ethical hotels across Britain, across Europe and into East Africa and Latin America in partnership with the local Methodist churches. All from one question at one meal. That question challenged the relationship between what they said and what they did. But they undertook the process of self-awareness themselves. We did not tell them what to do. Just asked why they weren't doing it yet!

To change values, you need to know what the values are in the first place. And this is where the faiths have the edge. About four-fifths of people around the world would say they follow a faith tradition… and these faith traditions have their values clearly set out. In sacred texts; in teachings; in practice; in their histories and their stories and legends; in the great charismatic figures such as saints and gurus.

So why has the environmental movement not bothered to work with them as key allies? Because there is still this ridiculous idea that the faiths are only concerned with how to get to Heaven. That is nonsense! The faiths are actually the largest and most consistent values groups on the planet and they are very much concerned about justice and kindness in this world. They have a great deal to say about schools and hospitals and care homes… in fact without their role in compassionate care, the world would be a sadder, badder place.

In the midst of the Second World War, when Britain's food supplies were threatened by the German attacks on the food convoys from America, the great Archbishop of Canterbury, William Temple, made a memorable statement. "We are three meals away from barbarism", he said. And anyone who has watched people fighting at petrol stations when supplies are briefly disrupted, or seen supermarkets emptied during a snowstorm, will know what that is.

But what if we flipped that? What if instead of three meals being the doorway to chaos and collapse we made the next three meals ones where we asked questions.

"How was this animal kept?"
"What sort of life did it have?"
"How was it killed?"
"How was the soil treated that grew these vegetables?"

"Who picked these fruits and how were they paid?"
"What was the packaging?"
"How compassionate is this whole meal?"

Because if we thought like this, and if these questions were asked in more restaurant kitchens, at school dinner tables (we have to have answers that we are not ashamed by), at conferences or on aeroplanes and trains and of course in each of our homes... then one meal at a time we can change the world.

Perhaps not always as swiftly and surely as the egg question changed the Methodists but making every meal a question, a challenge and a possibility.

ARC has worked with the World Wide Fund for Nature (WWF-UK) for many years and they have been great supporters and friends to us. But for at least 20 years we have often held our trustees' meeting at WWF-UK. And every time we have asked for vegetarian food only (and preferably organic and carefully sourced), as part of ARC's principles. And they have supplied, it but as an exception. I was delighted when we went last month to hear that they have now decided to go fully vegetarian for their visitor catering. One meal at a time.

# 22 The Sustainable Development Goals

## Challenge or opportunity?

*Hans R. Herren*

In September 2015, after nearly three years of deliberation, all countries agreed to a new sustainable development agenda, the Agenda 2030, the Sustainable Development Goals (SDGs), which are enshrined in 17 *universally* accepted Goals, 169 targets and 242 indicators. This is the logical continuation of the original efforts started in 1992 at the Sustainable Development Summit in Rio de Janeiro, where the Agenda 21, with the Millennium Development Goals (MDGs), was developed. This created the basis for a new development agenda that would advance the protection of the environment and natural resources while also looking at the social aspect of development.

The 17 SDGs are a major step towards achieving sustainable development across the globe and across its many dimensions; this is in contrast to the seven MDGs, which were confined to the developing countries. The 17 goals do add to the complexity of getting the development job done, but they are a necessity, given the need to cover all possible bases to assure that development is sustainable in the long run.

It very important to note that in the context of agriculture and the food system, the SDGs do include the main recommendation of the International Assessment of Agricultural Knowledge, Science and Technology for Development (IAASTD) (2009), which I co-chaired with Dr Judi Wakhungu, and which was commissioned by six United Nations (UN) Agencies and the World Bank at the Sustainable Development Summit in Johannesburg in 2002. The main statement made by the 400 authors from around the world was that "Business as usual is not an option" and that agriculture and the food system need a radical transformation towards agroecology and sustainable development. Sustainable development includes the environment, society and the economy, and is an integral part of the food system. Governments, and even its 59 signatories have done little since its approval in 2008. It's a great success of the over 150 non-governmental organisations (NGOs), led by the Biovision Foundation and the Millennium Institute, that they did manage to get the language of the IAASTD report into the Rio+20 Declaration and eventually into the SDGs.

SDG 2, Zero Hunger, is very central to the need to change course in agriculture and food systems. It is interesting and important to note how central Goal 2 also is to achieving all the other goals. This is unsurprising, given the centrality of food and nutrition security in sustainable and equitable development.

Why do we need to change course with our agriculture and food systems everywhere around the globe? On a global scale, we produce about twice the food needed; however, we grow the wrong kind in the wrong way, in the wrong places and often for the wrong uses, such as for biofuels or animal feed. We are dealing with the burden of malnutrition, hunger, micronutrient deficiencies, obesity and non-communicable diseases (NCDs). In addition we also have, as a result of the present agriculture practices, environmental unsustainability, that is, biodiversity losses; water pollution; soil degradation; greenhouse gas (GHG) emissions; unsustainable use of natural resources; low resilience; and social inequities, like poverty; disempowerment; and, not least, neglect of cultural values. All this calls for urgent action.

We do know what to do, how to do it and who should do it. We also know that it is feasible from a financial point of view. The evidence has been published in a number of post-IAASTD reports (Green Economy UNEP, 2011). What has been lacking is the political support, a change in agricultural and food system policies that would support the transformation process. A recent report from the International Panel of Experts on Sustainable Food Systems (IPES-Food) (IPES-Food 2016) has highlighted the blockages and offered ways to make the transformation. Despite what we know, the policymakers are still in the old green revolution mode, supporting an agriculture of the past, which contributes to climate change, produces empty calories, uses natural resources unsustainably and destroys the bases of agriculture. This kind of intensive agriculture results in soil degradation instead of soil enrichment, as would be the case if farmers around the globe embraced organic agriculture, agroecology, permaculture, or in more general terms, regenerative agricultural practices linked to a resource-saving food system.

The present food production systems have replaced the ecosystem services, which have supported food production over the aeons with nutrient and water cycling, pollination, pest control and broad genetic diversity constantly adapting to new environmental conditions. The current reductionist system is using pesticides, synthetic fertilisers and hybrid and genetically modified (GM) seeds, and is based on a productivist and short-termist view. The consequences are here to see, an agriculture that is bankrupt, dependent almost everywhere on subsidies or price support, epidemics of obesity, type 2 diabetes, malnutrition, hunger and poverty, not to mention the degradation of natural resources upon which agriculture depends, plus its contribution to climate change. A totally different approach could enable agriculture to be part of the solution to all these problems. We know better, and the time has come to act.

In the areas of overproduction and waste on a grand scale, such as the US, Europe and other industrial countries, we could do with more diversity in crop and livestock production, even with a small drop in production, which would easily be compensated for with increased resilience. There would still be more than enough by 2050 to nourish the population well. In the developing countries and countries in transition, production today is still low and can be easily doubled or tripled using organic agriculture practices, as the evidence from research in Africa and Asia has shown. This would then also suffice for a well-balanced and healthy nutrition of the growing population. Instead and unfortunately, major development agencies, foundations and many governments still promote the agricultural model of the past, the green revolution productivity model. Instead they could and should promote a modern, science-based and multifaceted agriculture as can be seen in today's organic agriculture, agroecology, permaculture and regenerative agriculture. Of course these practices do not fit the industrial, mega large-scale model, where the input providers are in control and dictate what is being grown when, where and how, and then also control the market. But so be it, change is needed and agribusiness will have to adapt or move out of the way. Consumers and citizens are already rightfully rebelling over the heavy multiple health burdens loaded on them by a food system that has been proven over and over to be more of a problem for food and nutrition security than a solution.

For change to happen, we need to find some key leverage points. One that is being heavily discussed and probably the most realistic, is true costing. This means that the price of food, just as with any other goods, should carry the full production costs as well the positive (minus costs) and negative externalities (plus costs) (see Peter Stevenson's chapter for more detail on this). With such a pricing system, sustainably produced food would be cheaper than the conventional and industrially produced food. Externalities are the loss of biodiversity, soil degradation, overuse of water, destruction of ecosystem services, health costs, water pollution, etc. On the positive side we have the benefits to ecosystem services from resource-caring production practices, better health through nutrient-rich and residue-free food, etc. This approach to transforming the food system is now being given serious attention by several projects such as TEEB-Ag from The United Nations Environment Programme (UNEP). (TEEB stands for The Economics of Ecosystems and Biodiversity.)

Fortunately, we now have a universally government-agreed policy framework to support the needed changes, the SDGs. We need to make the best use of them, make sure that new food system policies are developed in the SDG spirit of inclusiveness and full accountability. The SDGs open the door to a systemic and holistic approach to development and we need to make full use of these characteristics for the benefit of the food systems we want. There are many development partners now on board for making the needed changes a reality. The initial push from the French government at the Paris COP 21, with the "4 pour mille" initiative (www.4p1000.org/), shows that there is a movement to support agriculture's transformation from a carbon emitting and climate change promoting one to one that will help reverse climate change by not only being neutral but also absorbing excess $CO_2$. This while improving soil structure, water retention and nutrient cycling….. healthy soils, healthy agriculture, healthy planet and healthy people. This is where we need to go, what we can do, so let's do it now.

## References

Agriculture at a Crossroads, 2009. International assessment of agricultural knowledge, science and technology for development (IAASTD), 2009 Island Press.

Full references for the statements and documents listed can be found in the book: "How to nourish the world" by Hans R Herren, 2017 rü er & rub Sachbuchverlag GmbH, Zurich.

Green Economy, UNEP, 2011. http://drustage.unep.org/greeneconomy/resources/green-economy-report.

IPES-Food, 2016. From uniformity to diversity: a paradigm shift from industrial agriculture to diversified agroecological systems. International Panel of Experts on Sustainable Food systems. www.ipes-food.org/images/Reports/UniformityToDiversity_FullReport.pdf.

Island Press is a trademark of The Center for Resource Economics.

Library of Congress Cataloging-in-Publication data.

# 23 Rethinking our global food systems

*Jonathon Porritt*

## The productivist fantasy

The biggest problem with our global food system is that the people who were thinking about the future of food back in the 1960s and 1970s have educated a second generation of people who are thinking in exactly the same way! And they all suffer from a chronic productivist fantasy.

Their thinking is spectacularly simplistic: since there will be x billion more people on planet Earth, we will therefore need x billion more tonnes of food. It sometimes seems to me that this is the limit of their vision for what any new food system will look like in the future.

It's easy to poke fun at this fantasy, but this is a very dangerous voice, as they're very influential, both in government and with big funders. One can see their influence most clearly with a lot of the philanthropic money currently going into the future of food and nutrition. Indeed, it is these productivist fantasies which lie at the heart of why we're still finding it so difficult to come up with intelligent alternatives.

## The old ways

If you look at what is happening around the research agenda for the future of food and farming – which is a very good indicator of the likelihood of changing this system fast enough to make a big difference – 80% of the research in developed world countries is still mostly ignoring the phenomenon of accelerated climate change, both in home country production systems and as it impacts on imported food. It is simultaneously ignoring the fact that we can't go on using nitrogen-based and mined phosphorus as our principal source of nutrients and it is mostly ignoring what is happening to our soils – not least because of the excessive use of artificial fertilisers and chemicals – and totally ignoring energy balances. How many calories of energy do we need to produce a calorie of food? When I was growing up in the world of sustainability back in the 1970s, that was held to be the single most important relationship in terms of understanding what genuinely sustainable food production systems would need to look like. That has been completely eliminated from our discussions. That research agenda now is still dominant, despite what we know about the demands for producing, distributing and consuming foods very differently before 2030 – when the impacts of climate change have properly kicked in.

## Creating a better future

I'm hoping that people are beginning to grasp the inherent unsustainability of our current approaches to food production, and in particular, to the still dominant role of animal-based protein in the food system.

And this is where I have to touch on the question about population. It's part of my job as "an agent provocateur" to remind people that at the heart of many of today's sustainability dilemmas is the questions of how many heads (or stomachs in this case!) we have to accommodate the needs of on this finite planet.

It's this refusal to confront the question of population, amongst other things, which makes it an extremely ambitious target to move away from today's unsustainable food production and distribution systems to ones which will be genuinely sustainable within the course of the next 25–30 years.

However, I believe this process of change is already under way, and that it could now accelerate in ways that people currently find hard to believe. This is spearheaded not only by a lot of young people's awareness but also by the speed of innovation. Technology change is hugely important in this area. By way of comparison, it wouldn't be possible to talk about a genuinely sustainable energy world for nine billion people if it weren't for some of the new technologies coming forward around renewables, storage, smart grids and so on. In the same way, it's impossible to talk about genuinely sustainable and nutritious diets for nine billion people by 2050 without an equally compelling level of innovation in the food industry.

Unfortunately, a lot of people in that industry still have incredibly old-fashioned ideas. Their mindsets are still warped by the productivist fantasy that I referred to above, which means, for instance, that they've never really thought through the huge negative impacts of livestock production systems on any number of environmental problems today.

And I sometimes think that's true of a lot of environmentalists and conservationists. I'm constantly surprised by the number of people who care passionately about biodiversity, threatened species and ecosystems, who've never thought seriously about the connection between intensive livestock productions systems and those parts of the natural world that they care so passionately about. It may sound strange, but people love staying comfortably within their particular silo, and never want to look too broadly outside it!

## Opportunities beckon

But the really important story here is to think about opportunities for people who understand the world as it really is, rather than hark back to the world as it once might have been. And in that story, business has got a really important role to play. For me it's interesting to see that most of the big retailers, for instance, are developing some increasingly smart and focussed ideas about recalibrating the whole protein story for their consumers. But they're uncomfortable talking about it. They're not quite certain how far to go, for instance, when it comes to challenging people's meat consumption habits.

And maybe there really will be an opportunity, in this surreal post-Brexit world bearing down on us, in the UK to think about ways of transforming the debate. To put a new framework of sustainable nutrition at the heart of food and farming. Personally, I find it extremely difficult to believe anything that Government Ministers are saying at the moment, but we need to go on challenging them vociferously as to what they really think the future might entail. And let's hope our big environmental organisations play a really dynamic and creative part in that process. For me, it's pretty extraordinary that they are uncomfortable talking about population, and equally uncomfortable, it would seem, talking about the damage done through intensive animal production systems. That really can't continue.

Which means that I'm all the more delighted that Compassion in World Farming was able to bring together so many people at their Extinction and Livestock conference – and now, in this book – to share their insights into the problems associated with the livestock industry, and how we need to rethink radical alternatives to build a more sustainable future.

# 24 Sustainability in the agro-food sector

*Karl Falkenberg*

Some of my colleagues have written about the global impacts of our present agro-food system on climate change, biodiversity and health. Others have discussed animal welfare, the scientific assessments of animal feelings, bonding and intelligence and the ethical consequences for our relationship with animals. My own take on these issues comes from an assessment of the sustainability of our present agro-food sector.

The European Union (EU), which I have represented in a range of different functions over a period of 40 years, has been actively involved in the definition of the UN Sustainable Development Goals (SDGs), adopted by the UN General Assembly in September 2015. This process originally started in 1992 in Rio de Janeiro, with the first Earth Summit. Twenty years later, again in Rio, the UN held a further Conference on Sustainable Development and called for the negotiation of SDGs. It is these 17 well-defined goals, the SDGs, which were eventually adopted as a universal call to end poverty, protect the planet and ensure prosperity for all.

The EU and all its Member States not only signed up to these goals and committed to their implementation by 2030 but strongly insisted on their universal application. This therefore requires an assessment of present EU policies with regard to their sustainability. It also requires a better understanding of the notion of Sustainability, which brings together the economic, social and environmental consequences of human activities on Planet Earth. The continued rapid rise in world population (from a mere three billion in the 1950s, when I was born, to the seven billion of today and an expected ten billion in 2050) has highlighted the limits of our planet: human activity is responsible for the rapid loss of biodiversity, the acceleration of climate change and the profoundly uneven distribution of wealth and use of the planet's resources. The planet is not on a sustainable path, nor are the EU's own policies.

The agro-food sector is a good illustration of what has gone wrong in terms of sustainability.

We do feed more people than ever before, which certainly is a major achievement. But we begin to see the evidence that for all three aspects of sustainability we need to change: neither the economics, the social nor the environmental consequences of our current systems are sustainable.

Economically, agriculture employs fewer people, creates fewer, larger farms and contributes to the continued migration from rural to urban areas. It creates wealth in fewer and fewer hands, exacerbating income inequalities. The race towards cheap food not only cuts into the revenue of many farmers but is also partly responsible for the fact that 30% of the food we produce continues to be wasted.

Socially, in addition to the employment and income problems, our agro-food system contributes to growing health problems. Our Western diets produce cardiovascular problems, obesity and certain cancers, which are rapidly growing health issues, as Frank Hu has explained so well in Chapter 27. It also creates major concerns with antibiotic use in animal production, threatening resistance to antibiotics in bacteriological diseases for humans. Residues from fertilisers, pesticides and herbicides in our food can increase health risks like cancer. Growing numbers of our population are questioning the quality of standard food and beginning to change their diets.

Environmentally, our present agricultural practices create equally worrying trends. Biodiversity losses are particularly dramatic in agriculturally used areas. Recent studies quote a 70% reduction in insects, and bird populations continue on balance to shrink. Industrial animal husbandry leads to the nitrate pollution of surface waters, affecting not only our rivers but also the surrounding seas. The way we produce meat and milk with imported animal feeds is an additional environmental pressure on forestry ecosystems in the rest of the world and hence a substantial contribution to accelerated climate change. Agriculture is a major drain on our water resources and the quality of our soils. It is also one of the significant sources of air pollution in the form of fine particles, such as ammonia and nitrogen.

EU policies have been a major cause of the present situation. Agricultural policies have been aimed at increased productivity, favouring large-scale farming. These policies were motivated by a need to rebuild Europe's food self-sufficiency after the Second World War. The need for change has since been understood, but is resisted by agricultural federations, the agro-industry and the food sector. Despite several reforms, the bulk of agricultural subsidies still goes in the form of income support to the largest farms. The rural development fund recently created is beginning to help more sustainable forms of agriculture, based on the idea of public support for public goods, both environmental and social. In some sectors, such as wine and olive oil, producers have understood and accepted that less quantity with better quality results in better farm income. In most commodities, however, the race to cheap food continues. This is so despite the growing awareness that the advice to farmers that more fertiliser and pesticide use, more sophisticated seeds and more expensive farming machinery will create better farm incomes is not true because the additional investments tend to simply produce further declining commodity prices. There may be an alternative because those farmers who have resolutely turned their back on the existing model have discovered that lesser use of pesticides and fertilisers produces higher yields and better farm income. In the case of durum wheat production, farmers realised that insect pollination increases yields.

Organisations like Compassion in World Farming play an important awareness-raising role with regard to all these challenges. Their Extinction and Livestock conference has offered a real opportunity for an exchange of ideas. We are all consumers and as such we have a strong card to play. If consumers buy sustainable food, producers will shift to meet their demand. Large majorities of European consumers (over 70%) tell us in opinion polls that they are willing to pay more for healthier, sustainably produced food. However, the market realities do not reflect these numbers as organic, bio or fair trade products represent at best 5%–7% of sales. Unfortunately there is this huge gap between our awareness and our actions. One cause for this apparent incoherence may be the lack of trust in the large number of competing labels, making informed consumer choices a real challenge.

The SDGs, adopted at UN level, require more decisive actions to halt the raging environmental destruction and growing social injustice than what can be achieved through awareness raising alone. And we need change urgently! By 2030, we have committed ourselves through the SDGs to ending poverty and hunger, to changing our production and consumption patterns, to ending gender discrimination and reducing inequalities, leaving no one behind. We are committed to saving our oceans and our terrestrial ecosystems. We will need to change our policies and align them with the jointly identified and adopted sustainability goals. This will require courageous policymaking, which in our representative democracies relies on elected people. The issues raised at Compassion's conference need to be translated into political action. In addition to our consumer card, we have a political card to play: at all levels of democratic governance, we need to elect representatives into our Parliaments who pay more than lip service to the UN SDGs and are committed to urgently adopt sustainable policies. We need to hold our local and national governments and the EU itself to their promises made at UN level for a sustainable future!

The views expressed are strictly those of the author. They do not represent views or positions of the EU Commission.

# 25 Farming insects for food or feed

*Phil Brooke*

## Introduction

According to the United Nations Food and Agriculture Organisation (FAO), it is estimated that insects form part of the traditional diet of around two billion people (Van Huis, 2013). Over 1,900 kinds of insects have been reported as being used for food. These include the famous "Witchety grubs", a group of beetle larvae which are a favourite food of Australian Aborigines. It also includes a range of palm weevils whose larvae feed on felled sago palm trees. Some of the insects consumed are best known as pests. For example, where locust swarms are common, they often form part of the diet.

Most are harvested from the wild but, for example in Thailand, 20,000 farmers rear crickets. In South East Asia, some kinds of palm tree are deliberately felled to encourage colonisation by palm weevils.

These traditional uses already provide an important source of nutrition and livelihoods. It is now being suggested that, to feed a rising population demanding higher quantities of animal protein, we should expand the use of insects both as human food and as feed for farm animals across the globe.

## Efficient use of protein

We grow more than enough protein today in the form of beans, cereals, vegetables and grazing livestock to feed the current and likely future population.[1] Unfortunately, food isn't distributed equitably and hundreds of millions of people remain malnourished. Meanwhile, a third of the cereals and over 90% of the soya protein produced globally is fed to animals. In the process of converting plant protein to meat, fish or eggs, 60% or more of that protein is wasted, and only 3%–40% is converted into human-edible food (Fry et al., 2018).

If we want to ensure that there is enough protein for everyone without ploughing up more land and increasing agrichemical use, then the priority is to use protein more efficiently. It is most efficient to:

1 Feed human-edible food to people
2 Feed food that is not edible for people to animals

3 Feed food that is not edible for farm animals to insects, or other invertebrates, such as worms or woodlice, to transform it into feed for farm animals
4 Use any remaining substrate in a biodigester to produce methane (usually for energy, but it could be used to grow methanogenic bacteria for animal feed)
5 Compost any remains and return to the land.

Insects can play a role in this efficiently if they are fed on wastes to feed people (wherever people can be persuaded to eat them) or where they are converting wastes which are not edible for other farm animals into animal feed (wherever this can be both safe and legal). Feeding insects on human-edible feed, or producing animal feed from wastes which could be fed directly to other farm animals, is less likely to be efficient.

Protein production could also be increased efficiently through the fermentation of microbes to produce human-edible food (e.g. mycoprotein, yeast or algae) or animal feed (bacteria, yeast or algae).

## Farming insects for human food

It is argued that insects convert their feed into their body tissue more efficiently than warm-blooded animals such as chickens and pigs. They don't need to expend energy keeping warm and their bodies contain less indigestible material (e.g. bones, feathers, etc.). This means there should be:

- A lower food conversion ratio (FCR), which measures the efficiency of converting insect feed measured by weight into live insect.[2]
- A lower FCR (edible weight) which measures the efficiency of converting feed into the edible part/s of the insect.
- A more efficient protein conversion ratio which measures the proportion of protein in the feed which makes its way into the edible insect.
- A lower calorie conversion ratio which measures the proportion of calories in the feed which makes its way into the edible insect.

Farmed insects are commonly reared on poultry feed. According to research, the FCR (edible weight) for growing crickets (*Acheta domesticus*) on poultry meal was 1.7. This compared with FCRs (edible weight) of 2.3 for carp, 2.4–4.2 for chicken, 14 for pork and 31.7 for beef (Lundy & Parella, 2015).

In short, converting poultry feed into insects appears to be more efficient than rearing fish, chicken or red meat. This is partly because they don't need to expend energy keeping warm and their bodies contain less indigestible material (e.g. bones, feathers, etc.).

However, to grow animals as "efficiently" as this requires a high-quality feed. The poultry feed used in this case contained mainly maize and soya grain products. For many people the purpose of consuming animal products is to obtain high-quality protein, but to obtain insect protein this efficiently requires a high-protein diet. A better measure of efficiency would therefore be the protein conversion rate rather than FCR.

The protein conversion efficiency for crickets is still only 35% (Lundy & Parella, 2015). Black soldier fly larvae can do better with an efficiency of 43%–55% (Oonincx et al., 2015). This compares with 21%–40% for chicken, 14%–28% for farmed fish and crustaceans, 9%–21% for pig meat and 3%–13% for beef. So, rearing insects is generally less inefficient, but around half to two-thirds of the protein is still lost in conversion.[3]

Eating insects instead of fish, pork or beef would make your diet more efficient; however, it would be more efficient still to eat some of the maize and/or soya directly.

A better answer might be to feed the insects on food waste. For example, the FCR of black soldier fly larvae fed on waste material can be as low as 1.4 (Oonincx & De Boer, 2015).

To get such an efficient FCR, they had to feed the fly larvae on a highly concentrated diet. In this case it was based on food waste with a high nutritional quality, but if they reduced the food quality, the FCR rose to 2.6. If they reared species more commonly considered as candidates for human food, such as mealworms or crickets, then the FCR rose to 3.8–19.1 according to species and food source.

Nevertheless, if the insects are reared on food waste, it is a gain to food security since they are consuming food that people wouldn't consume. Food waste which can be legally fed to insects in the European Union (EU) could also be fed to fish, poultry or pigs, but the figures in Table 25.1 show that insects are likely to convert the waste more efficiently, provided that they were directly consumed by people.

*Table 25.1* Conversion efficiencies of insects compared with other farm animals

| *Animal* | *Protein conversion efficiency* | *Calorie conversion efficiency* | *Food conversion ratio (FCR)* | *Notes* |
|---|---|---|---|---|
| Beef | 3%–13% | 3%–7% | 8–14 | Poor converters of grain (FCR 8), but commonly fed on forage which people cannot eat (so high FCR doesn't imply inefficiency) |
| Pork | 9%–21% | 9%–16% | 3.1–6.5 | |
| Chicken | 21%–40% | 12%–27% | 1.9–2.5 | High protein feed required for efficient FCR |
| Farmed fish and crustaceans generally | 14%–28% | 6%–25% | 1.6 | Fish generally fed on very concentrated high protein feeds, so FCR exaggerates efficiency |
| Yellow mealworm | 22%–50% | | 3.8–19.1 | These figures were based on a range of experimental diets based on wastes and other food manufacturing byproducts (e.g. yeast) as well as a specially formulated control diet (e.g. poultry feed for black soldier fly larvae) |
| House cricket | 23%–41% | | 2.3–6.1 | |
| Black soldier fly larvae | 43%–55% | | 1.4–2.6 | |
| Argentinean cockroach | 51%–87% | | 1.5–2.7 | |

Most figures except insects – Fry et al. (2018). Most insect figures Oonincx et al. (2015). Higher figures for pork and beef – Lundy and Parella (2015).

However, in the West there is a reluctance to consume insects. Culturally, it might be easier to persuade people in the developed world to eat plant-based foods than to eat insects. Indeed, it is for this reason that many of the proponents of insect farming see it primarily as a way of producing animal feed, not human food. But, once you add a trophic level to the food chain, you almost certainly lose any efficiency advantage as we shall discuss in the next section.

## Farming insects for animal feed

Chickens and pigs naturally eat invertebrates, including insects, as part of their omnivorous diet. So do freshwater fish; marine fish eat invertebrates, although not insects since, whilst there are some aquatic insects in freshwater, there are very few species found in seawater.

Insects are being considered as feed for a range of food security and environmental reasons:

- As an alternative protein source to mitigate the environmental costs of producing soya or fish meal.
- As an alternative protein source to fish meal for farmed fish given that this is a finite resource and that capture of these fish depletes the oceanic food chain.
- To enrich the diets of omnivorous species such as poultry.

The food security and sustainability benefits depend on what the insects are fed on:

1. They could be fed on human-edible foods such as cereals and soya. This adds a new level of inefficiency. The insects will use or excrete part of the energy and protein in the food, so there will be less in the feed than if the soya and cereals were used directly. Even more gets used or lost by the animal which eats the insects. This is clearly not an efficient way of producing animal feed. It could also increase, rather than reduce, demand for soya and fish meal if these are part of the insect diet.
2. They could be fed on food wastes such as bakery wastes, surplus vegetables or any other food left over from the direct human food chain. European legislation permits a restricted list of insect species to be fed on a restricted group of food wastes (Commission Regulation, 2017/893). All of these are food wastes which are currently permitted as feeds for other farm animals. It is likely to be more efficient to feed these wastes directly to animals to produce meat, milk or eggs, rather than, once again, wasting part of the food value by passing them through insects on the way. Feeding these wastes to insects to produce feed for farm animals is likely, at best, to provide marginal benefit.
3. Insects could be fed on substrates like manure which are outside the normal human food chain.[4] If this were permitted, there would be obvious advantages for food security and efficient land use. However, this raises

questions of food safety and public acceptability which are unlikely to be overcome, at least in the near future. The EU legislation referred to in the last paragraph specifically excludes the use of such substrates as insect feed for farm animals.

Perhaps a better use for wastes which are not suitable for human or animal consumption is to use them for biofuels. It seems perverse that we make such fuel out of human-edible food whilst attempting to make safe dubious sources of food which would make perfectly good biofuel.

Research is ongoing into developing waste streams that may be suitable for growing insects for feed. I suspect that any feed that can be made safe for insects could also be made safe for other farm animals to eat directly and that the latter would be normally be a more efficient use of the feed source. But, time will tell.

## Use of insects in the diets of fish

Carnivorous fish need a high concentration of protein in their diets (Henry et al., 2015) as is provided in fish meal. They also need diets high in long-chain omega-3 fatty acids such as docosahexaenoic acid (DHA). In natural conditions they would obtain these by eating other fish and aquatic invertebrates.

The production of fish meal and fish oil for consumption by farm animals is controversial because of problems of over-fishing and the depletion of oceanic food chains. Fish meal and oil from sustainable sources are available, but the quantities are finite and are already fully exploited. The production of even "sustainable" fish meal and oil also raises very substantial welfare issues for the 450 billion or more fish that are caught for this purpose each year, without any practical method of humane slaughter being available, to supply the fish farming industry (Mood & Brooke, 2012).

The industry has substituted vegetable proteins and oils to facilitate the expansion of carnivorous fish farming which is otherwise near its sustainable limits. However, due to problems with anti-nutritional factors which can result in digestive problems and questions of palatability (carnivorous fish like to eat fish), there have been limits to the levels of substitution possible (Henry et al., 2015).

Insects are high, sometimes very high, in protein and can also be used as a substitute for fish meal, for example in fish diets (Lock et al., 2016). In so far as the production of insect meal reduces the pressure on forage fisheries, this might be considered a good thing. However, the main factor limiting the production of carnivorous fish is fish oil, not fish meal. Fish oil contains essential long-chain fatty acids such as DHA, originally manufactured by algae at the beginning of aquatic, but not terrestrial, food chains. These fatty acids are not generally found in significant quantities in terrestrial insects (Henry et al., 2015). You can increase levels by adding these fatty acids to the insect diet, for example by feeding them on fish waste, but this would defeat its own purpose.

All of this raises the question of whether it makes any food-security sense to farm carnivorous fish in the first place. We don't farm carnivorous land animals and for good reason. At every stage in a food chain, both energy and protein are lost. It is for this reason that it is nearly always most efficient to eat plant food (unless the animals are consuming food which we could not eat). Eating animals which have eaten human-edible food is wasteful. Eating animals which have eaten animals which have eaten human-edible food is more wasteful still.

There are other reasons to question whether rearing insects for feed to facilitate an increase in the production of carnivorous species of fish would be a positive step:

1 Increases in the rearing of carnivorous marine fish have led to problems with build-up of disease and parasites affecting the welfare of both farmed and wild fish; with fish wastes polluting the sea floor and surrounding waters the impact of medicines and other chemical discharges on marine life; with escapes of farmed fish risking the genetic diversity of wild stocks; the impact of the capture of cleaner fish such as wrasse; with the impact on marine mammals such as the control of fish predators such as seals; (Scottish Parliament, 2018).
2 Fish farming has developed rapidly without a full understanding of the physiological and, in particular, the ethological needs of the fish. It is not well understood what impact confinement to a sea-cage will have on the well-being of species which may migrate over 1,000 miles in the wild.
3 Possible means of solving the problems in no 1 above, for example by keeping the fish even more intensively in recirculation systems which remove them altogether from their natural environment – factory farming in other words – risks further problems for fish welfare and public acceptability.
4 Practical, safe and legal means of rearing insects as fish feed have the potential to increase the production of farmed fish, but the process is not likely to be an efficient use of food resources. Any expansion brings environmental and welfare risks.

## Feeding insects to chickens or pigs

Insect meal has been suggested as an alternative to soya or fish meal in pig and poultry rations. Pigs and chickens are omnivores and insects and other invertebrates are a natural part of their diet and trials show that insect meal can be successfully included in their diets (Veldkamp & Bosch, 2015). It is also a food these animals would enjoy.

Pigs and chickens can be fed on a vegetarian diet, so it makes no sense to feed them on insects fed on poultry feed. Pigs and chickens can also be fed directly on food waste. If the food waste is fed first to insects, they can concentrate the protein component, thus reducing the amount of soya used in the feed; however, in concentrating the feed, they are wasting much of the energy, and some of the protein also, in the process. It would still be better to feed food waste directly to pigs and chickens, if necessary keeping more robust breeds which are better adapted to diets which are lower in protein.

As discussed earlier, if it ever becomes both safe and legal for pigs and chickens to consume insects which have eaten feed such as manures that cannot be fed to animals, this could transform the situation.

## Economics of insect production

Insect protein production can be expensive. According to one study published in 2011, a kilogram of mealworm protein cost 31.70 euros compared to 1.91 for fish meal and 0.62 for soybean meal (Meuwissen, 2011).

No doubt these figures will come down, but it is not uncommon for new food sources to be heralded but not to pass the economic test. I remember that Imperial Chemical Industries Ltd (ICI) produced a bacterial protein called "Pruteen" from methanol (derived chemically from natural gas) back in the 1970s (Braude et al., 1977). This could have been used in place of soya beans in pig and poultry diets, saving much *Cerrado* from the plough, but soya was cheaper.

It remains to be seen whether insect production will prove economic, except for niche markets such as pet food and wild bird feed. Competing in the market as a protein source will also be the descendants of Pruteen. Fermentation processes already produce mycoprotein (Quorn), yeast and synthetic amino acids. Micro algae are also grown as health foods and as vegan fish oil alternatives. More of these later.

## Insect sentiency and welfare

In their 2013 report for FAO, Van Huis et al. include a section on insect welfare. They argue, or quote arguments, to the effect that:

- Insects should be given sufficient space, in line with their ethological needs.
- They should be killed using methods which reduce suffering.
- That while little is known about their capacity for pain and discomfort, they should be given the benefit of the doubt whilst conclusive proof is gathered.

In relation to the ethological requirements of insects, they note that locusts in captivity are gregarious and that mealworms have a tendency to cluster.

The animal kingdom is divided into a number of groups, each called a phylum. Insects belong to a massive phylum called the Arthropoda. Apart from the insects, other classes of arthropod include the Crustacea (e.g. crabs and lobsters) and the Arachnids (e.g. spiders). Arthropods generally have segmented bodies and jointed limbs.

There are a wide group of different orders of insect, mostly found in two divisions. One division changes gradually at each moult, showing incomplete metamorphosis. This includes the Orthoptera (grasshoppers, crickets, locusts), termites and the Odonata (dragonflies and damselflies). Wings commonly emerge at the final moult.

The second division of insects goes through complete metamorphosis with a larvae going through an immobile pupal stage where a wingless larva metamorphoses into a more highly developed and usually winged adult. This includes the Diptera (flies), Coleoptera (beetles) and Hymenoptera (bees, wasps, ants).

Amongst invertebrates, the greatest evidence for sentiency has been found amongst cephalopod molluscs (Octopus, squid and cuttlefish) and decapod crustaceans (crabs, lobsters, prawns) (Broom, 2013). Broom argues that some insects such as bees and ants have quite a high level of cognitive ability and probably "assessment awareness" and that there is a case for some degree of protection for insects.

The evidence for sentiency in insects includes:

1. Insects exhibit nociception. This is the system, exhibited by a wide range of vertebrates and invertebrates, which enables the body to sense and respond to a wide range of injurious stimuli which cause damage such as heat, mechanical or chemical injury. Interestingly, some of the genes which control nociception are the same as in mammals (Neely et al., 2011).

   In many species, the process of nociception can be controlled by opioids such as morphine, which commonly provides pain relief. A drug called naloxone is sometimes used medicinally as an opioid blocker (it prevents morphine from working). Honeybees, praying mantis and crickets are known to produce opioids during defensive reactions and to have opioid receptors that are blocked by naloxone (e.g. Groening et al., 2017). Attempts to show that opioids work as analgesics in insects have not, as yet, produced positive results.
2. Honeybees make decisions which imply "cognitive bias", that is, experience may make them "optimistic" or "pessimistic". Sweet food can increase positive emotions and improve negative mood in human adults, through dopamine release.

   Bees that have just had sucrose are more likely to approach a stimulus that is halfway between a previously trained positive and negative stimulus, that is, they may be more optimistic as to the outcome. The sucrose did not affect their activity level or make them more likely to approach totally novel stimuli, only ones that were intermediate between the trained ones.

   Bees given a dopamine antagonist took longer to approach the intermediate stimulus suggesting that dopamine release was affecting their behaviour.

3 Ants teach each other in a process called tandem running (Richardson et al., 2007). In an experimental set-up, their home is destroyed. Scouts head off to find a new suitable site. When a scout finds one, she returns to the nest and shows another one – who follows her in tandem, regularly stopping after each step to learn coordinates for the route, until they reach the new site. Both then return and each teaches another ant the route. This is repeated till a fair proportion know the way whereupon those who know the way carry the rest of the group (this is three times faster than tandem running).

   Unlike many other examples of animal teaching, this involves some evaluation or feedback. The teacher ant takes a step forward and waits for a cue to continue further. If, experimentally, the learner is removed, the teacher will wait a minute or so before giving up and looking for another learner. If, however, the learner has been disabled by removing an antennae so that they take longer to give the cue to carry on, when they are removed experimentally the teacher ant will wait for a shorter time before giving up and looking for another. The teachers are less patient with slower learners.

4 Matabele ants go on frequent raiding parties to termite mounds to predate on termites. Some of their number are commonly injured during the raiding party. Injured ants are commonly rescued and carried back by their comrades (Frank et al., 2017). After rescue, the ants are tended and treated, a process in which their wounds are licked. This has been shown experimentally to reduce the risk of death by infection (Frank et al., 2018).

5 Ants have passed the "mirror test" (Cammaerts & Cammaerts, 2015). If a blue mark is painted on their clypeus, a place on the head important in ant recognition, they try to clean it. Responding to a reflection as being "of oneself" in this way is a rare accomplishment in the animal kingdom, and the first example in an invertebrate.

The assessment of the sentience of insects is clearly in its infancy. Furthermore, all the research discussed above has been carried out on adult insects and none on larval stages. Most of the insects being considered for conversion to animal feed, for example blackfly, housefly and mealworm, are all likely to be harvested towards the end of the larval stage (it is different with locusts and crickets which are adults of insects that show incomplete metamorphosis). Adults are of course required for breeding.

The larval nervous system is less developed than that of the adult. Along with other arthropods, insects have a decentralised nervous system. There is a brain in the head end but also a series of centres in the nerve cord called ganglia which also store information and coordinate the muscles within that segment.

In fruit flies, dipteran flies called drosophila (which are related to black soldier flies and houseflies), the larva emerges with a fully functional primary brain when it hatches. This is followed by development of secondary tissue which lies dormant awaiting metamorphosis. During pupation, the primary brain is restructured and the secondary tissue develops into additional brain structures only found in the adult (Hartenstein et al., 2008). There is also further growth and development in the nerve cord.

So, are insects sentient? If so, do they become sentient at the larval stage or does this only apply to the adult? The truth is, we don't know. More research is required, especially for the larval stage.

However, if they are sentient, the scope for suffering is massive. Fed on waste, black soldier fly larvae may weigh around 0.15 g (St-Hilaire et al., 2007). This means that there can be over 5,000 to the kilo or 5 million to the tonne. To match the 20 million-plus tonnes of fish caught each year to produce fish meal to feed farm animals would require 100 trillion of these animals.

For this reason, one must agree with the arguments presented by FAO, that insects including larvae should be given the benefit of the doubt (Van Huis et al., 2013). If insects are reared for food or feed they should be reared, transported and killed humanely. This requires an understanding of their ethological needs as well as physical requirements for health and welfare. It also requires scientifically determined methods of humane killing, not just a general expectation that heating, freezing or grinding would provide a quick death. This is likely not so.

## Other impacts of farming on insect welfare

Insect pests are killed in very large numbers, usually through the use of insecticides. Many insecticides are nerve poisons, some of them derived from nerve toxins originally developed as chemical weapons. If insects are sentient, then insecticides will cause very poor welfare to very large numbers of them.

Insecticides are used primarily on arable, fruit and nut crops; sometimes in forestry and almost never if at all on pasture (although grazing animals themselves may be treated). Chemical insecticides are not permitted in organic farming.

Reducing the amount of arable crops required by reducing the amount of human-edible food fed to animals including insects would reduce the requirement for insecticide use. Less arable land would be needed and it could be farmed less intensively. Reducing food waste and feeding wastes to animals including insects would also help reduce the amount of arable land required to grow crops.

Rotational farming with fewer inputs, as practised by organic farmers, would also reduce the requirement for insecticide use.

## Other alternative novel foods

In addition to rearing insects, there are other strategies for producing the protein and calories required to feed people more efficiently.

## Eat more plants

We already grow enough grains, fruits, nuts and vegetables to supply the human population with sufficient protein and calories and to have plenty over.

In practice much of any surplus is lost due to food waste and the wasteful practice of feeding human-edible food to animals. When grain is fed to animals, most of the calories and protein are metabolised by the animal and lost in conversion. Further is lost to the human food chain due to incorporation into tissue which is not eaten such as bones, intestines, etc. Eating protein directly is more efficient than feeding the protein to animals.

The simplest most natural way of doing this is to eat more plant foods: grains, nuts, seeds, fruits and vegetables.

There is also an increasing plethora of foods designed to replace meat, milk and eggs using a range of protein sources including soya, wheat gluten, peas, nuts, etc.

## Feed animals on food that we cannot eat

Feeding animals on grass and food waste which people couldn't eat can increase the amount of food available to people. This can include feeding insects for human consumption.

## Growing microbial protein

Unlike animals, microbes such as bacteria, algae, yeast and other fungi can make their own protein if they are given a supply of energy and suitable nutrients (including nitrate and sulphur compounds which are needed to make the amino acids which constitute protein).

Microbes can be grown in fermenters using sugars grown in agriculture. These include nutritional yeast, mycoprotein (Quorn), bacteria and heterotrophic algae. The latter are currently grown to provide oils containing the long-chain fatty acids eicosapentaenoic acid (EPA) and DHA as a vegan alternative to fish oil capsules and as an alternative to fish oil in fish-feed. Algal protein is also produced at the same time.

Whereas rearing animals for food wastes two-thirds or more of the protein consumed, these microbes create protein without consuming it. They do, however, inevitably consume calories in the process in order to be compliant with the laws of thermodynamics. This means you do have to grow human-edible food – for example sugar or wheat starch – so it could be argued that this is still wasteful. Why not grow beans which are high in protein in the first place?

Growing beans is certainly a good idea, but growing wheat or maize to produce mycoprotein could produce more protein per hectare. First, the average yields of cereals per hectare are much higher than beans. Second, the yield of protein per hectare from wheat – in the form of gluten which can be used to make the seitan often used in faux chicken – is almost as high as the yield of protein per hectare from soya beans.[5] The starch left over can be converted to sugar to grow mycoprotein maximising protein yield per acre.

Similar principles could be applied to growing other microbial proteins. This includes the growth of yeasts which have been genetically engineered to produce particular animal proteins, such as:

- Milk proteins which can be used to produce milk alternatives such as "Perfect Day" (www.perfectdayfoods.com/).
- Egg albumin as an alternative for recipes which require egg white (www.clarafoods.com/).
- Plant "haemoglobin", as used in the "Impossible Burger".

In all these cases, these technologies increase rather than reduce the production of protein per hectare at a cost of some of the energy that could have been consumed in the crops they are produced from. It remains most efficient to obtain much of one's protein directly from grains and vegetables that provide one's carbohydrate energy requirements, but this is an efficient way of topping that protein up.

Some microbes can be grown entirely without the need to input human-edible food:

- Phototrophic micro algae can be grown in ponds using energy from sunlight. Not only do they have a high protein content, they can also produce long-chain fatty acids such as DHA. Oil companies such as Esso are researching into the use of micro algae as a source of fuel for transportation. If this ever becomes economic, it may produce a substantial amount of algal protein as a byproduct.
- Methanogenic bacteria can produce protein using methane as an energy source. One source of such protein has been approved by the EU for the production of feed for aquaculture and is being developed commercially.[6] Feed like this can be produced using minimal land and water resources, although the carbon footprint is high unless biogas is used as the methane source (Cumberlege et al., 2016).

An alternative to feeding wastes of uncertain safety to insects would be to put the waste in a biodigester, using the methane produced to grow microbial protein.

## Growing animal cells in culture

A key efficiency advantage of rearing insects for human consumption is that the whole animal is commonly eaten, perhaps apart from wings in adult insects so that protein isn't wasted on non-edible parts. Another means of obtaining the same objective would be to grow cells in culture. There are a range of projects aiming to produce beef, pork, chicken or fish products in a test tube (www.new-harvest.org/portfolio).

It is speculatively estimated that compared with rearing farm animals, cultured meat will require "approximately 7–45% lower energy use (only poultry has lower energy use), 78–96% lower GHG emissions, 99% lower land use, and 82–96% lower water use depending on the product compared" (Tuomisto et al., 2011). Presumably this does not include the land, water and energy required to grow the food nutrients required to feed the culture.

Inevitably there will be protein and energy loss here, but the expectation is that it will be relatively low and might be expected to compare with the efficiency of insect production.

A big gain to food security would be to use new technologies, for example the production of algae, to supply the nutrients needed to grow the cell culture. As always, it would be more efficient to eat the algae directly.

## Making more efficient use of grasses and other leaves

When I was growing up in the 1970s, I remember leaf protein being put forward as a food of the future. If you extract juice from grass, or other similar green leaves, and heat the juice you extract, it can precipitate protein that you can filter out. It was reputed to be very high in protein, with a good amino acid balance and essential fatty acids (including short-chain omega-3), vitamins and minerals, and to be a healthy and tasteless protein which could be added to stews, etc. It was also proposed that it could be fed to pigs and chickens with the fibrous residue used as cattle feed.

I bought a juice extractor to test this out on grass from my lawn, and it worked, though the protein coagulate produced was extremely bitter, and I didn't repeat the experiment. Wheat grass juice is also sometimes sold as a health food – it is less bitter but still gastronomically challenging.

The idea is being considered again (Santamaría-Fernández et al., 2017) as a means of providing protein for pigs and poultry. In this case, the proteins are precipitated by lactic-acid fermentation rather than heat. A range of leafy crops were tested including clover and alfalfa. Amino acid composition is comparable to soybeans, although methionine levels are higher, making this an attractive prospect for organic farming. It could also work well in any rotational farming system, producing protein from crops on which the use of insecticides and fungicides is not normally required.

## Conclusion

The world is not short of the protein and energy required to feed people, but poverty and inequitable distribution mean that much of it is wasted feeding animals or biofuels, so hundreds of millions of people end up hungry or malnourished.

Insects harvested from the wild or farmed on a small scale play an important part in the diets and livelihoods of many people in developing and emerging countries. Insect farming is considered by some to be an important means of increasing protein production to ensure that more people are fed.

Whilst there is some scope for the expansion of insect production, its potential for increasing protein supplies is limited and other sources may be more promising. Rearing insects for human consumption on feed that is not human edible would increase protein supply. It would also be beneficial for food security and environmental protection if people ate such insects as an alternative to factory farmed vertebrate animals or fish meal-fed prawns. However, it may be challenging to spread the habit of eating insects widely.

In principle, it would also be possible to rear insects for animal feed on wastes that could not be fed safely or legally to farm animals. Free-range pigs and chickens would certainly welcome such an addition to their diet. However, there remain safety, and in some cases legal, obstacles to using such substrates in the human food chain. The EU, for example, bans the use of feed for insects that could not legally be fed directly to farm animals.

Insects are often raised on poultry feed, the main ingredients of which (cereals and soya) are human edible. It is argued that insects use feed more efficiently than other farmed animals, and for some species with some diets this is potentially true. It would probably be more efficient to eat a black soldier fly burger than a pork chop or a chicken breast. However, it would be more efficient still to eat the cereals and soya directly.

Insects could be reared for feed on wastes that are suitable for consumption directly by farm animals. However, even if reared efficiently, nearly half the protein is lost in that first conversion. It is likely to be more efficient to feed those wastes to farm animals directly.

Meanwhile, there are other alternatives which can efficiently increase the amount of protein available for people to eat. These include:

- Eating more plant-based foods and meat alternatives.
- Growing microbial protein including mycoprotein, yeast, bacteria and algae for both food and feed.
- Extracting leaf protein from grasses, clovers and other leaves to feed monogastric animals such as pigs and chickens and possibly directly to people.

Research may well develop new feed streams that can produce insect protein more productively, but fermentation technologies have vastly more potential to increase protein production for food and feed efficiently than insect rearing. These technologies can also be scaled up without creating factory farms for insects.

The development of insects as a source of feed may depend on economics. Production of insects is currently expensive and the production of microbial protein may prove to be cheaper. Time will tell.

Are insects sentient? There is some evidence for this, at least in adult insects. Further research may well establish a great deal more evidence. In the meantime, they should be given the benefit of the doubt. If insect farming takes off, the numbers produced is likely to be in the trillions, possibly tens or hundreds of trillions, so the potential for suffering could be very high. Therefore, effective steps should be taken to ensure that suffering is avoided during rearing, transportation (if any) and killing.

On this basis, on a modest scale, the development of insect farming could have the potential to be sustainable to produce food for those people who want to eat them and to enrich the diets of free-range farm animals. The danger is that it could be developed on a much larger scale, using human-edible feedstocks, with factory farmed insects producing feed for factory farmed pigs and poultry. This would create suffering on a massive scale for the pigs and chickens and, potentially, on an astronomical scale for the insects. All of this to waste grain which could have fed the hungry.

Alternatively, we could eat more plant-based food and distribute the earth's ample protein supplies more equitably. We could develop fermentation technologies to increase supplies of protein for both food and feed. We could raise our sheep and cattle on pastures which maintain or improve soil fertility, and our humanely raised pigs, chickens and insects could be raised in moderate numbers on other foods which we cannot eat.

This would enable us to avoid consigning more of the earth's habitats to the plough, to reduce the use of insecticides and to farm in more nature-friendly ways.

## Notes

1 A quick calculation using The Food and Agriculture Organization Corporate Statistical Database (FAOSTAT) figures for crop production and human population, and United States Department of Agriculture (USDA) figures for protein content of foods, suggests that the world produces over 40 g of soya protein per person per day as well as 35 g of maize protein, 25 g of wheat protein and 20 g of rice protein. Ignoring all other foods, these four plant foods produce together more than twice the normal human requirement of 50–60 g of protein per day. It is true that maize and wheat are low in the amino acid lysine, but soya is high in this. If one counts additional protein from other sources including grass-based livestock and vegetables, it is clear that we produce well over twice the daily requirement for human protein.

However, it is not equitably distributed and most of this is lost as a result of feeding much of the wheat and most of the maize and soya to animals. Analysis of other FAOSTAT figures shows that people eat 49 g of plant protein per day, on average, including 31 g from cereals, 7 g from vegetables and 7 g from pulses (including peanuts).

Total average protein consumption averages 81 g, including animal protein.

2 FCR is a measure of the efficiency of food conversion, equal to food-in divided by food out. The higher the FCR, the less efficient the conversion. For example, if a mealworm needs to eat 3 g of feed to put on 1 g of weight, the FCR is 3.

3 Oonincz et al. do quote protein conversion ratios from 51% to 87% for the Argentinian cockroach, depending on feedstuff. Like ruminants, this creature

harbours bacteria in the digestive system which can produce protein using sources of energy and nitrogen. The higher level of protein retention was achieved using a diet high in fat as a source of the energy. Unfortunately, as with ruminants, the bacteria also produce methane which is a greenhouse gas. This species is commonly fed to insectivorous pets, but is not currently permitted by the EU as a food source for people or farm animals.

4 Some manures are used to grow mushrooms.

5 Admittedly, the quality of protein is lower in wheat due to its low content of the amino acid lysine. Soya is high in lysine and contains a good balance of amino acids but is slightly low in another amino acid – methionine. Combining the two together – wheat is quite high in methionine – produces high-quality protein.

6 www.feedkind.com/what-is-feedkind/.

## References

Braude, R., Hosking, Z.D., Mitchell, K.G., Plonka, S. and Sambrook, I.E., 1977. Pruteen, a new source of protein for growing pigs. I. Metabolic experiment: utilization of nitrogen. *Livestock Production Science*, *4*(1), pp. 79–89.

Broom, D.M. 2013. The welfare of invertebrate animals such as insects, spiders, 3 snails and worms. In Animal Suffering: From Science to Law, International 4 Symposium, ed. Kemp, T. A. van der and Lachance, M., (pp. 135–152). Paris: Éditions Yvon Blais.

Cammaerts Tricot, M.C. and Cammaerts, R., 2015. Are ants (Hymenoptera, Formicidae) capable of self recognition? *Journal of science*, *5*(7), pp. 521–532.

Commission Regulation (EU), 2017/893 of 24 May 2017 amending Annexes I and IV to Regulation (EC) No 999/2001 of the European Parliament and of the Council and Annexes X, XIV and XV to Commission Regulation (EU) No 142/2011 as regards the provisions on processed animal protein. http://eur-lex.europa.eu/legal-content/EN/TXT/PDF/?uri=CELEX:32017R0893&qid=1501019934927&from=EN.

Cumberlege, T., Blenkinsopp, T. and Clark, J., 2016. Assessment of environmental impact of FeedKind protein. *Carbon Trust*. URL www. carbontrust. com/media/672719/calysta-feedkind.pdf.

Frank, E.T., Schmitt, T., Hovestadt, T., Mitesser, O., Stiegler, J. and Linsenmair, K.E., 2017. Saving the injured: Rescue behavior in the termite-hunting ant Megaponera analis. *Science Advances*, *3*(4), p. e1602187.

Frank, E.T., Wehrhahn, M. and Linsenmair, K.E., 2018. Wound treatment and selective help in a termite-hunting ant. Proc. R. Soc. B, 285(1872), p. 20172457.

Fry, J.P., Mailloux, N.A., Love, D.C., Milli, M.C. and Cao, L., 2018. Feed conversion efficiency in aquaculture: do we measure it correctly? *Environmental Research Letters*, *13*(2), p. 024017.

Groening, J., Venini, D. and Srinivasan, M.V., 2017. In search of evidence for the experience of pain in honeybees: A self-administration study. *Scientific Reports*, 7, p. 45825.

Hartenstein, V., Spindler, S., Pereanu, W. and Fung, S., 2008. The development of the Drosophila larval brain. In *Brain development in Drosophila melanogaster* (pp. 1–31). Springer, New York.

Henry, M., Gasco, L., Piccolo, G. and Fountoulaki, E., 2015. Review on the use of insects in the diet of farmed fish: past and future. *Animal Feed Science and Technology*, *203*, pp. 1–22.

Lock, E.R., Arsiwalla, T. and Waagbø, R., 2016. Insect larvae meal as an alternative source of nutrients in the diet of Atlantic salmon (Salmo salar) postsmolt. *Aquaculture Nutrition*, *22*(6), pp. 1202–1213.

Lundy, M.E. and Parrella, M.P., 2015. Crickets are not a free lunch: protein capture from scalable organic side-streams via high-density populations of Acheta domesticus. *PloS One*, *10*(4), p. e0118785.

Meuwissen, P., 2011. Insecten als nieuwe eiwitbron. *Een scenarioverkenning van de marktkansen. ZLTO-projecten.'s Hertogenbosch, The Netherlands,* quoted in Veldkamp et al. 2012.

Mood, A. and Brooke, P., 2012. Estimating the number of farmed fish killed in global aquaculture each year. *Fishcount.org.uk*, p. 1.

Neely, G.G., Keene, A.C., Duchek, P., Chang, E.C., Wang, Q.P., Aksoy, Y.A., Rosenzweig, M., Costigan, M., Woolf, C.J., Garrity, P.A. and Penninger, J.M., 2011. TrpA1 regulates thermal nociception in Drosophila. *PloS One*, *6*(8), p. e24343.

Oonincx, D.G., Van Broekhoven, S., Van Huis, A. and van Loon, J.J., 2015. Feed conversion, survival and development, and composition of four insect species on diets composed of food by-products. *PLoS One*, 10(12), p. e0144601.

Regulation (EC) No 1069/2009 of the European Parliament and of the Council of 21 October 2009 laying down health rules as regards animal by-products and derived products not intended for human consumption and repealing Regulation (EC) No 1774/2002 (Animal by-products Regulation. http://eur-lex.europa.eu/legal-content/EN/TXT/PDF/?uri=CELEX:32009R1069&from=EN.

Richardson, T.O., Sleeman, P.A., McNamara, J.M., Houston, A.I. and Franks, N.R., 2007. Teaching with evaluation in ants. *Current Biology*, *17*(17), pp. 1520–1526.

Santamaría-Fernández, M., Molinuevo-Salces, B., Kiel, P., Steenfeldt, S., Uellendahl, H. and Lübeck, M., 2017. Lactic acid fermentation for refining proteins from green crops and obtaining a high quality feed product for monogastric animals. *Journal of Cleaner Production*, *162*, pp. 875–881.

St-Hilaire, S., Cranfill, K., McGuire, M.A., Mosley, E.E., Tomberlin, J.K., Newton, L., Sealey, W., Sheppard, C. and Irving, S., 2007. Fish offal recycling by the black soldier fly produces a foodstuff high in omega-3 fatty acids. *Journal of the World Aquaculture Society*, *38*(2), pp. 309–313.

Scottish Parliament, 2018. Review of the environmental impacts of salmon farming in Scotland. http://www.parliament.scot/S5_Rural/Inquiries/20180125_SAMS_Review_of_Environmental_Impact_of_Salmon_Farming_-_Report.pdf

Tuomisto, H.L. and Teixeira de Mattos, M.J., 2011. Environmental impacts of cultured meat production. *Environmental Science & Technology*, *45*(14), pp. 6117–6123.

Van Huis, A., Van Itterbeeck, J., Klunder, H., Mertens, E., Halloran, A., Muir, G. and Vantomme, P., 2013. *Edible insects: future prospects for food and feed security* (No. 171). Food and Agriculture Organization of the United Nations (FAO).

Veldkamp, T. and Bosch, G., 2015. Insects: a protein-rich feed ingredient in pig and poultry diets. *Animal Frontiers*, *5*(2), pp. 45–50.

# 26 Hopeful signposts to a plant-based future

*Carol McKenna*

## Introduction

I'm a sci-fi fan and will share, quite unashamedly, that I often enjoy movies that almost every critic roundly condemns. One such is the 2008 adaptation of *The Day the Earth Stood Still*, starring Keanu Reeves as Klaatu, an alien who's come to save the Earth. There's an amazing quote in the film that is often on my mind when I'm working towards ending factory farming in favour of regenerative agriculture. It comes during a conversation between Klaatu and Professor Barnhardt, a scientist, played by John Cleese, who is trying to dissuade him from destroying the human race so that the Earth and its non-human life can survive. When Klaatu reveals that his people were forced to evolve because their sun was dying, Barnhardt replies,

> Well that is where we are. You say we are on the brink of destruction and you are right. But it is only on the brink that people find the will to change. Only at the precipice do we evolve.

It seems to me that humankind is very close to the precipice. We live as if we have Planets B, C and D within commuting distance to satisfy our needs and wants including our ever-increasing demands for energy, transport, food, entertainment and experiences. We've already exceeded important planetary boundaries and have damaged the very life systems of our planet, as Katherine Richardson explains in the opening chapter to this book.

2050 is fast approaching, with its predicted human population of nine or ten billion people. What will life be like for our children and grandchildren if we don't change our ways? We're already using nearly half the world's usable land to produce food (Searchinger et al., 2013). Wildlife is already disappearing 1,000 times faster than the normal rate of extinction (De Vos. et al., 2014). How much habitat will be left in future for the animals with whom we share the planet, unless we change? Just one example is that there are some 20,000 African lions and 500 Asiatic lions left. 50 years ago, there were 450,000 lions. Studies show that every time the human population increases by one billion, the lion population decreases by 50% (Langin, 2014). We're facing not only a world without iconic animals such as lions, tigers and elephants but also a world without even the smallest of creatures, like the bees and worms so vital to pollination and soil health (Lymbery, 2017).

How does one find hope when faced with a planetary emergency on this scale? Well, after becoming even more concerned whilst reading Steven Emmott's *10 Billion* (Emmott, S., 2013), which left me with huge fears for my son's future, I've found my hope with the recognition that the transformation of our food and farming systems is the way forward.

Reducing meat and dairy consumption in high consuming populations, whilst managing production and consumption globally, has the potential to enable governments to meet the Paris Climate Change Targets and several of the Sustainable Development Goals (SDGs). Indeed it will be almost impossible to avoid exceeding the 2°C threshold if we don't substantially reduce meat and dairy consumption in the developed world (Bojana, Benton and Clark et al., 2015). Recent analysis published in *Science* reveals that meat, aquaculture, eggs and dairy use 83% of the world's farmland and contribute 56%–58% of food's different emissions, despite providing only 37% of our protein and 18% of our calories (Poore and Nemecek, 2018). The researchers concluded that dietary change towards vegetable proteins has transformative potential with respect to vastly reducing food's land use and greenhouse gas (GHG) emissions as well as other beneficial environmental impacts (Table 26.1).

The evidence for action to reduce meat and dairy consumption and to transform our food systems is clear.

The question is how to make this happen. My aim with this chapter is to draw attention to some of the practical initiatives that demonstrate that change is possible, and that it's happening right now.

*Table 26.1* Environmental impacts that would arise by moving from current diets to a diet that excludes animal products (2010 reference year)

| *Environmental impact of food* | *Reduction arising by moving from current diets to a diet that excludes animal products* |
|---|---|
| Land use | 3.1 billion hectares reduction; this is a 76% reduction including a 19% reduction in arable land |
| GHG emissions | 6.6 billion metric tons of $CO_2$eq; this is a 49% reduction |
| Acidification | 50% reduction |
| Eutrophication | 49% reduction |
| Scarcity-weighted freshwater withdrawals | 19% reduction |

Source: Poore and Nemecek (2018).

Leading food policy experts, Dr Pamela Mason, an independent nutritionist, and Professor Tim Lang of the Centre for Food Policy at City University of London have advanced the very compelling argument of an "$SDG^2$" strategy for public policy. That is to say, national and local Sustainable Dietary Guidelines to support delivery of the SDGs because only low impact diets will stop current food systems from undermining our future (Mason and Lang, 2017).

Sustainable diets are,

> those diets with low environmental impacts which contribute to food and nutrition security and to healthy life for present and future generations. Sustainable diets are protective and respectful of biodiversity and ecosystems, culturally acceptable, accessible, economically fair and affordable; nutritionally adequate, safe and healthy; while optimizing natural and human resources.
>
> (FAO, 2010)

In order for sustainable diets to be realised, Mason and Lang suggest a phased transition of our food systems over the next few decades with goals being set for all sectors under the headings of food quality, health, environment, culture, economy and governance. In a paper written for Friends of the Earth proposing the $SDG^2$ approach, Lang advises,

> If national government will not engage, then civil society, city authorities and purpose-led businesses must step into the vacuum, while we maintain pressure on government to come to its senses. Other levels of and actors in food governance can effectively promote and implement such guidelines through policies and practices including choice editing, sustainable marketing and positive, value-shifting messages about health, pleasure, convenience, social interaction, taste, and ethical policies.
>
> (Lang, 2017)

As this chapter explores, many actors are doing exactly that.

## Government action on sustainable dietary guidelines

To date relatively few countries, most notably Brazil, Germany, the Netherlands, Qatar, Sweden and the UK, have explicitly included sustainability within their national dietary guidelines (Health Council of the Netherlands, 2015; Gonzalez Fischer and Garnet, 2016; Public Health England, 2016a; Mason and Lang, 2017; Netherlands Nutrition Centre, 2017). The actions of these countries signpost the way forward for others, as all propose limits on meat consumption and call for more plant-based diets, as shown in Table 26.2.

*Table 26.2* Healthy and sustainable dietary guidelines

| *Brazil* | *Germany* | *The Netherlands* | *Qatar* | *The UK* |
|---|---|---|---|---|
| • Healthy diets derive from socially and environmentally sustainable systems<br>• Eat foods mainly of plant origin<br>• Make natural, minimally processed foods the basis of your diet<br>• Try to restrict the amount of red meat | • Choose mainly plant-based foods<br>• Enjoy five portions of fruit and vegetables daily, as fresh as possible<br>• Eat meat, sausages and eggs in moderation – no more than 300–600 g of meat and sausages per week | • Eat more plant-based and less animal-based food<br>• Limit consumption of red meat, particularly processed meat<br>• Maximum of 500 g of meat a week<br>• Eat plenty of wholegrain products, fruit and vegetables. At least 200 g each of vegetables and fruit a day | • Emphasises a plant-based diet, including vegetables, fruit, whole grain cereals, legumes, nuts and seeds<br>• Aim for 3–5 servings of vegetables and 2–4 of fruits every day<br>• Limit meat (500 g per week)<br>• Avoid processed meats | • Eat at least five portions of fruit and vegetables every day<br>• Base meals on starchy carbohydrates, preferably whole grain<br>• The protein section starts with beans and pulses<br>• Eat less red and processed meats |

Sources: Gonzalez Fischer and Garnett (2016), Health Council of the Netherlands (2015), Netherlands Nutrition Centre (2017) and Public Health England (2016a, 2016b).

Brazil, with its estimated average retail meat consumption of 215 g per person per day, is one of the largest meat-eating countries in the world (OECD, 2017), so its dietary guidelines provide a particularly heartening example. They affirm that "healthy diets derive from socially and environmentally sustainable food systems" and attest that "dietary recommendations need to take into account the impact of the means of production and distribution of food on social justice and environmental integrity" (Ministry of Health of Brazil, 2015).

Importantly, they state,

> choosing diets based on a variety of foods of plant origin with sparing amounts of foods of animal origin implies the choice of a food system that is relatively equitable, and less stressful to the physical environment, for animals and biodiversity in general.

They also note that the country's traditional cuisines are based on plant foods with sparing consumption of meat, adding,

> Reduced consumption and thus production of animal foods will reduce emissions of the greenhouse gases responsible for global warming, of deforestation caused by creation of new grazing areas for cattle, and of intensive use of water. It will also reduce the number of intensive

animal production systems, which are particularly harmful to the environment.

When deciding on updates to its nutritional guidelines the UK's Public Health England collaborated with the Carbon Trust to determine the sustainability implications for any changes (Public Health England, 2016a). Its 2016 Eatwell Guide emphasises the food groups that are more sustainable. It's noteworthy that the protein section now starts with beans and pulses rather than meat, with the specific intention of highlighting non-meat protein sources (Public Health England, 2016b).

## National dietary guidelines: less meat, more plants

2016 saw media coverage around the world of China's plans to cut meat consumption by 50% because of climate change, with claims that this would transform Chinese agriculture (The Guardian, 2016). The coverage was initiated by new dietary guidelines, which recommend cutting the weekly intake of livestock and poultry to 280–525 g or 40–75 g per day (Chinese Nutrition Society, 2016). In 2016 China's meat consumption was calculated at 136 g/day. A limit of 75 g/day would represent a reduction of nearly one half (45%), with a limit of 40 g/day, equating to a two-thirds reduction (OECD, 2017). The call for reduced meat consumption in the guidelines has been attributed to the Chinese government's concern about increasing meat consumption alongside rising levels of obesity and non-communicable diseases (Chatham House, 2017).

Other governments have called for reduced meat consumption in dietary guidelines, as outlined in Table 26.3, which shows 13 countries that are amongst the world's top meat eaters, excluding the European Union (EU). Of the 13 shown, the good news is that 12 have addressed the meat consumption issue either explicitly or implicitly. Furthermore, 9 of the 13 have advocated more plant-based foods, although some have done so only implicitly. Nevertheless these guidelines provide a starting point for environmental and health campaigners, as do the guidelines of several of the largest EU meat eaters, with Germany, for example, recommending a limit of 300–600 g of meat and sausage a week or 43–86 g per day (German Nutrition Society, 2013). This compares with probable current 178–241 g consumption a day (Netherlands Environmental Agency PBL, 2011; Statista, 2018). If the current consumption figures are accurate then this is a recommendation of a 50% reduction.

Italy's guidelines describe a serving of meat as 70 g, with two portions a day being recommended (Italian National Research Institute on Food and Nutrition, 2003). If Italians were to follow this guideline the limit of 140 g meat/day would bring about a reduction of about 15%–45% compared to the current estimated consumption level of 165–250 g/day (Netherlands Environmental Agency PBL, 2011; Statista, 2018).

*Table 26.3* Top meat eating countries and dietary advice on eating less meat, excluding EU

| *Country* | *Indicative published figures on meat consumption in retail weight, g per day, reported by OECD, 2017*[1] | *Recommendations on reducing meat* | *Recommends increasing plant-based foods* |
|---|---|---|---|
| Australia | 259 | Yes.<br>"Many Australian men would benefit from eating less meat"<br>*Maximum of 65 g/day limit for all of red meat*<br>*Eat less processed meat* | Implicit<br>*Plant-based meat alternatives listed with meat as proteins* |
| USA | 270 | Yes.<br>Advises lower meat intakes characteristic of healthy diet and lower CVD | No |
| Israel | 216 | No | No |
| Argentina | 243 | Only implicit<br>*Advises combining legumes and cereals to replace meat*<br>*Describes daily meat portion as size of palm of hand*<br>*Meat and fish shown in small part of recommended food plate chart* | Yes<br>*Five a day reduces risk of obesity, colon cancer and CVD* |
| Uruguay | 222 | Yes<br>*Reduce consumption of processed/cured meats and cold cuts*<br>*Very small slice of food plate chart shows meat and fish* | Yes<br>*Incorporate vegetables and fruits into all meals* |
| Brazil | 215 | Yes – see Table 26.1 | Yes – see Table 26.1 |
| New Zealand | 198 | Yes<br>*Links colon cancer to eating 71 g+/day of red meat and to processed meat*<br>*Limit processed meat*<br>*One serving of meat = two slices* | Yes<br>*Legumes and nuts listed equally with meat; states add more plant foods to make meat go further* |

(*Continued*)

*Table 26.3* (Continued)

| *Country* | *Indicative published figures on meat consumption in retail weight, g per day, reported by OECD, 2017*[1] | *Recommendations on reducing meat* | *Recommends increasing plant-based foods* |
|---|---|---|---|
| Chile | 197 | Only implicit<br>*Emphasises non-meat example menu items in healthy diet guide* | Yes<br>*Eat legumes at least twice a week without mixing with meat* |
| Canada | 192 | Only implicit<br>*Have meat alternatives often* | Implicit<br>*Emphasises vegetables and grains more than meat* |
| Malaysia | 150 | Only implicit<br>*Limit foods high in fat, moderate amounts of meat, use nuts and seeds to replace meat* | Yes<br>*Eat plenty of fruits and vegetables every day* |
| South Africa | 138 | Only implicit<br>*Adding small amounts of animal products to plant-based diet can yield health improvements*<br>One serving of cooked meat or fish is 85 g | Implicit<br>*Eat plenty of fruit and vegetables every day*<br>*Eat dry beans, split peas, lentils and soya regularly* |
| Republic of Korea | 153 | Yes<br>*Restrict fatty meats* | Implicit<br>*Eat vegetables and fruit every day and a lot of whole grains* |
| China | 138 | Yes<br>*Cut weekly intake of livestock and poultry to 280–525 g or 40–75 g per day* | Yes<br>*Advises vegetables and fruits are the main components of a balanced diet. Soy products to be eaten regularly* |

Note: Assumptions behind published figures on meat consumption vary widely: for example, whether available carcase weight, retail weight or actual consumption inside or outside the home are considered.

## Food businesses leading the way

Many influential, conventional food companies have adopted sustainability goals, including, for example, reducing energy, water, deforestation, GHGs, poverty, working to promote healthy eating, good global nutrition and improving worker rights, etc. (Sustainable Brands, 2018). However, few as yet are making links between their sustainability goals and the need to reduce consumption of meat and dairy products, with notable exceptions including global Italian pasta giant Barilla (Barilla, 2017), Compass Group US (Compass, 2016) and, of course, companies promoting meat alternatives. For example, Impossible Burger advises on its website, "Because we use 0% cows, the Impossible Burger uses a fraction of the Earth's natural resources".

Guided by its vision statement "Good for you, good for the planet", Barilla has founded a think tank on global food issues, the Center for Food & Nutrition Foundation (BCFN). This contributes to the development of healthier and more sustainable diets through a number of initiatives aimed at achieving the United Nations (UN) SDGs, including the Nutritional and Environmental Double Pyramid (Barilla, 2017), the Milan Protocol (Barilla, 2015) and the Eating Planet publication (Barilla, 2016).

Barilla's Double Pyramid (Figure 26.1) offers a dietary advice model that warrants global adoption because it clearly shows that foods with lower environmental impacts (fruit, vegetables) are also those recommended for their health benefits whilst demonstrating that foods with high environmental impacts (dairy products, meat) are those that should be consumed in moderation because of their negative health impacts. In a further demonstration of leadership Barilla reports that by 2020 it aims to offer only products in the bottom half of the environmental/food pyramid, that is, vegetables, fruits, grains and legumes.

Multinational food service company Compass Group, which operates in some 50 countries, serving some four billion meals a year, has a significant track record of promoting plant-based food, including working with partners, such as the Humane Society of the United States (HSUS), on flexitarian campaigns (Compass Group, 2017, 2018). In 2015 Compass Group announced a partnership with Hampton Creek Foods (now known as Just Inc.), which manufactures egg-free mayonnaise, cookies and dressings. Within six months that partnership had eliminated 1.2 million eggs from the supply chain (Compass Group, 2017). Importantly, Compass Group US's Envision 2020 plan includes reducing red meat purchases by 30% compared to 2014 and increasing plant-based items (Compass Group, 2016). It would be good to see Compass Group adopting similar policies in all the countries in which it operates.

Nowadays it seems that not a day goes by without media coverage of a new alternative to meat or a company launching more plant-based foods. Just a few notable examples follow.

In 2015 IKEA introduced plant-based alternatives to its meatballs in its stores worldwide. in its stores worldwide. The company serves food to some 660 million people a year with its People & Planet Positive sustainability strategy (IKEA, 2014) pledging to take a lead in a sustainable and healthy diet, where meat is included as a treat. In May 2018 IKEA announced plans

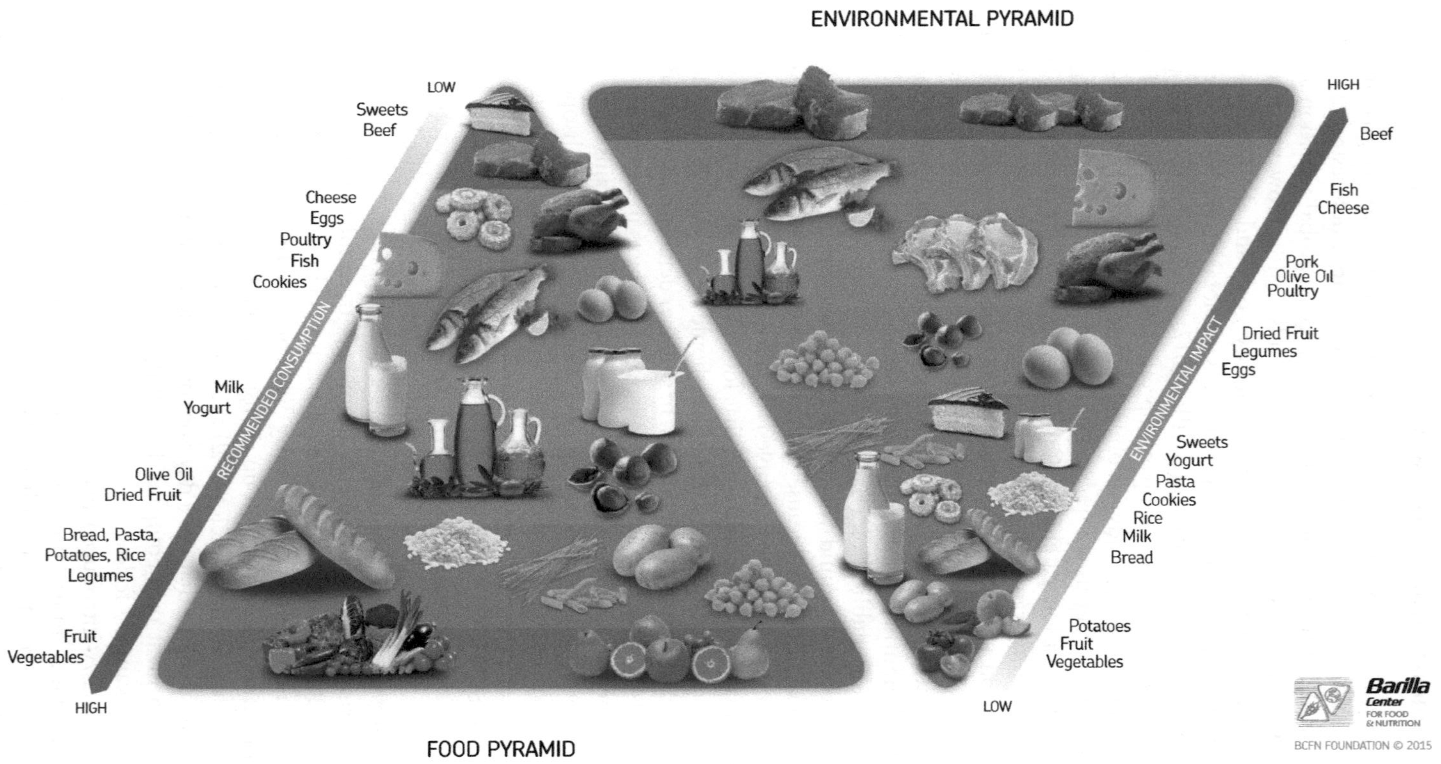

*Figure 26.1* The double pyramid.
*Source*: Barilla Centre for Food and Nutrition Foundation 2015.

to introduce a vegan hot dog across its European stores and to the US by 2019, advising that 12% of its Bistro customers reported choosing vegan or vegetarian options (Plantbased News, 2018).

In an unprecedented move by a retailer, 2016 saw Swedish supermarket chain Co-op launching a "Dear Meat" advertising campaign, explaining the reasons why, "even though most humans love meat, we have to eat less as eating as much as we do isn't sustainable". The print and TV-based ad offered practical help for people wanting delicious, easy-to-cook and kid-friendly vegetarian food, including recipes on social media (Forsman & Bodenfors, 2016).

2016 also saw Pret a Manger, which has outlets in 19 cities internationally, launching its "Not just for Veggies" campaign, described by CEO Clive Schlee as his promised "call to arms" marketing campaign for veggie food (Schlee, 2016). The "Not just for Veggies" concept is that everyone should try to make veggie choices more often. The year before, Schlee had invited customer views on how to respond to the increased demand for vegetarian foods, noting in his blog,

> There are good reasons to eat less meat (I'm trying!). Aside from the animal welfare arguments, the UN says that the single most important step an individual can take to reduce global warming is to adopt a meat and dairy-free diet.
>
> (Schlee, 2015)

The company expanded its meat-free options and for the month of June that year converted its existing Soho, London shop to vegetarian only. Its aim was to challenge its chefs and to consult customers on the new menu items. The campaign was so successful and demand for the veggie and vegan options is so great that Pret now has three vegetarian-only outlets in London and has expanded the campaign globally, launching, for example, more plant-based choices to its US stores in 2017 (Cision Newswire, 2017; Eating Better, 2017).

2017 saw Tesco, the world's third largest retailer operating in 11 countries, appointing pioneering chef Derek Sarno as its Director of Plant Based Innovation. This is an incredible development that sets a precedent for large retailers globally. Tesco and Sarno developed a Wicked Kitchen range of vegan ready meals, sandwiches and wraps that was launched across 600 UK stores in January 2018. May 2018 saw Tesco announcing that it would be launching a "breakthrough" vegan steak from Dutch manufacturer Vivera across 400 of its UK stores (Plantbased News, 2018). The retailer is also expected to be the first in the UK to offer the Beyond Meat Burger (The Grocer, 2018).

As highlighted by World Wildlife Fund (WWF)'s Glyn Davies in Chapter 30, Sodexo, a major food service company, has partnered with WWF-UK on developing a "Green & Lean" range, which includes meals that have had a certain percentage of meat substituted with pulses and vegetables. The UK has seen flexitarian sausages and mince containing a mix of meat with vegetables and/or pulses introduced by a number of retailers (Eating Better, 2017).

Some very exciting work involving Sainsbury's and the World Resource Institute's Better Buying Lab is exploring how language in product descriptions can be used to increase sales of plant-based items. Trials run in Sainsbury's stores have found that renaming products can increase sales considerably. The research findings are to be published during 2018 (Pers. Comm, 2018). Research by Stanford University also found that giving indulgent descriptions to vegetable dishes increased vegetable consumption (Turnwald, Boles, Crum, 2017). Others working in the plant-based field have recommended moving away from terms such as "vegan" and "vegetarian" to increase sales. Jennifer Pardoe, founder of PB2B, a product development agency specialising in the plant-based sector, advises clients,

> We need to see product branding and identities appealing directly to a meat eater's expected aesthetic and language. So they don't, for one second, consciously or subconsciously think those products as being only exclusive to vegans and vegetarians. Words like "vegan" or "vegetarian" should not be the headliner. Taste should be the headliner. Seth Tibbott, the founder of Tofurky always reminds us of the phrase "Taste is king, price is queen and everything else is positioning".
>
> (Pers. Comm, 2018)

The Sustainable Restaurant Association (SRA) launched an exciting initiative in 2018. The SRA's global *One Planet Plate* campaign aims to put sustainability on the menu and to help diners to use the power of their appetite wisely (One Planet Plate, 2018). Every chef is encouraged to add his or her sustainable special to the menu, making it easy for customers to choose a sustainable and ethical option. The SRA's sustainability framework criteria include urging restaurants to increase the proportion of vegetable-led dishes on the menu to combat environmental damage and to purchase high welfare meat and dairy products (SRA, 2017).

All of these initiatives are examples of actions that could be taken up by other food service operators, caterers, suppliers, manufacturers, chefs, restaurants and retailers. The market push for more and more businesses to move towards alternative proteins is becoming almost overwhelming, particularly in view of the number of investors using their considerable influence, as highlighted by Rosie Wardle, Bruce Friedrich and Michael Pellman Rowland in their chapters. As Jennifer Pardoe so succinctly explains, "Everyone in the food industry worth their salt knows plant-based is the dietary shift of the century" (PB2B, 2018).

## Sustainable diets for sustainable cities

Today more than half of the world's population lives in urban areas. The UN predicts that by 2030 urban areas will house 60% of people globally and considers city engagement to be crucial to delivery of the SDGs (United Nations, 2016).

Cities are taking action on food: for example, 167 cities have signed the Milan Urban Food Pact, which includes a commitment for cities to develop sustainable dietary guidelines (Milan Urban Food Pact, 2015).

Some cities have developed food policies that include the aim of reducing consumption of meat and dairy on sustainability grounds, for example, Brighton & Hove, which offers resources to support others wishing to follow (Brighton & Hove Food Partnership, 2012).

The City of Malmö's Policy for Sustainable Development and Food sets a good example for the world. The city has adopted the Eat S.M.A.R.T. model developed by the Institute of Public Health in Stockholm, which has five key recommendations including reducing meat consumption, increasing the amount of organic produce and carefully choosing the right sort of meat and vegetables from an environmental and health perspective (City of Malmö, 2010).

There appears to be tremendous scope for civil society and business engagement with cities and some organisations are spearheading action. In the UK the Sustainable Food Cities Award programme, run by the Soil Association, Sustain and Food Matters, encourages cities to take action on a range of key food issues including promoting healthy and sustainable food to the public (Sustainable Food Cities, 2018). Sustainable Food Cities also runs campaigns encouraging cities to engage on specific issues with 2018's theme being Veg Cities.

City climate change strategies provide excellent entry points for campaigning for healthy, humane and sustainable diets. Manchester's climate change strategy, for example, commits the city to becoming carbon-neutral by 2050 (Manchester Climate Change Agency, 2016a). The city's implementation plan nominates the University of Manchester and the Manchester Food Board to investigate pathways to assist citizens to adopt more sustainable and healthy food habits, including the potential to reduce meat consumption to sustainable levels (Manchester Climate Change Agency, 2016b).

## Civil society initiatives

Two of the best-known initiatives encouraging people to eat less meat are Meatless Mondays and Meat Free Monday. Launched in 2003, in association with the John Hopkins Bloomberg School of Public Health, Meatless Monday now operates in over 40 countries worldwide and has been taken up by at least 200 schools and 150 universities internationally (Meatless Mondays, 2018a and 2018b). In 2009 Paul, Mary and Stella McCartney launched Meat Free Monday, attracting support around the world including from celebrities, businesses, schools and universities (Meat Free Monday, 2018). It's interesting to note that the origin of Meatless Monday dates back to the First World War, when the US Food Administration introduced the concept alongside "Wheatless Wednesday" (Meatless Monday, 2018c).

World Meat Free Week is another initiative gaining more and more support (World Meat Free Week, 2018). Its 2018 partners include many non-governmental organisations (NGOs) and businesses, such as the previously mentioned Compass Group and Quorn, which now markets its over 100 mycoprotein products in 18 countries (Quorn, 2018).

Other notable initiatives providing models for change include the Eating Better alliance, HSUS's Forward Food Programme, the Soil Association's Food for Life Served Here awards programme, Friends of the Earth/Responsible Purchasing Network (RPN)'s work with cities, CreatureKind's pledge and Compassion in World Farming's Friendly Food Award and call for a new global agreement to replace factory farming with a regenerative food system, all of which are now briefly outlined.

## Eating better – a model for global change

Launched in 2013 Eating Better is a broad-based UK alliance of over 50 organisations spanning environmental, public health, animal welfare, social justice, research, professional and responsible producer interests. The alliance aims to raise awareness of the need for people to eat less and better meat and dairy for the sake of human health, the environment and animal welfare, and to promote practical solutions for policymakers, food businesses and changing behaviour. Its principles for eating less and better appear in Table 26.4.

The alliance surveys the British public with respect to their meat consumption, brings organisations together on particular projects and issues, produces reports for policymakers and food businesses, trials campaigns and acts to amplify the work of its individual members. The alliance has produced an excellent set of recommendations, summarised in Table 26.5, for food businesses interested in reducing their offer of meat and dairy products and encouraging more plant-based eating. These appeared in a report that also showcased the progress being made by 20 companies (Eating Better, 2017).

Eating Better's work appears to be making headway. Its latest public opinion survey found 44% of the British public willing or already committed to cutting down on – or cutting out – their meat consumption (YouGov for Eating Better, 2017). As Sue Dibb, the then Executive Director said at the Extinction and Livestock conference, the alliance would like to encourage the same approach in other countries, regions or cities as collaboration builds a more powerful voice for change (Dibb, 2017).

It is important to mention that amongst Eating Better's founding members were Compassion in World Farming, WWF-UK and Friends of the Earth. The organisations had collaborated for some years and had come to believe that there was a need for more organisations to engage on sustainable diets and meat reduction. Compassion had been a pioneer on the issue: in 2004 it published its groundbreaking *The Global Benefits of Eating Less Meat* report with a foreword by Jonathon Porritt (see Chapter 23 calling for a transformation of attitudes and behaviour towards meat-eating (CIWF, 2004).

*Table 26.4* Eight principles for choosing less and better (Eating Better, 2018)

1. **Choose better for the climate** means shifting the balance of our diets towards more plant-based foods, including plant-based sources of protein such as beans and pulses while eating less meat of all types. Choosing meat from "pasture-fed" animals can help lock carbon into the soil, but this only makes sense in the context of consuming considerably less overall.
2. **Choose better for animals** means choosing meat and dairy from well-managed production systems that enable natural behaviour, support good health and provide a natural diet. The simplest way of doing this is choosing products with a credible animal welfare certification.
3. **Choose better for nature** means choosing livestock products that have a diet based around local food sources and home-grown feedstuffs.
4. **Choose better for feeding the world fairly** means shifting diets away from meat and dairy overconsumption in high consuming individuals and countries. A "livestock on left-overs" approach would see lower levels of production and consumption, with animals reared in foraging systems or fed on crop byproducts and food waste/surplus. Ruminants can be kept on grasslands, where this brings additional benefits.
5. **Choose better for health** means shifting towards more plant-based diets would have health benefits for the majority of the population.
6. **Choose better for responsible antibiotic use** means choosing products that require minimal antibiotic use in their production. In practice, this means avoiding products produced intensively.
7. **Choose better for cutting waste** means valuing meat as a precious resource, making the most of each carcase and reducing the amount of wasted edible food.
8. **Choose better for livelihoods** means choosing meat and dairy from smaller scale, higher standard producers. Choosing meat and dairy with a known provenance can reconnect producers and their customers such as through farm shops, box schemes, farmers markets and independent butchers.

*Table 26.5* Recommendations for Food Business Action (Eating Better, 2017)

| | |
|---|---|
| Restaurants, fast food chains and the food service sector | • Develop menus to provide a greater choice of less meat/more plant- based dishes<br>• Introduce more exciting flavours and ingredients within existing plant-based products to attract a larger number of customers<br>• Consider describing plant-based dishes without using the words vegetarian/vegan<br>• Position lower meat and meat-free options at the top of menu at beginning/front of self-service counters<br>• Serve small portions of "better meat" e.g. free-range, organic, pasture-reared, and market as a distinctive speciality<br>• Ensure chefs are trained in plant-based cooking and recipe development<br>• Make plant-rich dishes/menu items affordable |

(*Continued*)

*Table 26.5* (Continued)

| | |
|---|---|
| Food product manufacturers and retailers | • Reformulate existing products to provide a greater ratio of plant protein<br>• Invest in or collaborate with companies offering plant-based meat substitutes<br>• Introduce more exciting flavours and ingredients within existing plant-based products to attract a larger number of customers<br>• Make plant-rich products available to purchase at an affordable price for customers<br>• Consider price promotions and placement within store to encourage trials and repeat purchasing |
| Engaging consumers | • Brand-led campaigns, highlighting the nutrition and sustainability benefits of a plant-based diet with less and better meat<br>• Collaborations with related brands or retailers<br>• Promoting campaigns e.g. Meat Free Monday World Meat Free Day, veg sandwiches during British Sandwich Week<br>• Promoting cuisines and recipes where the veg is the "star" of the meal<br>• Helping people with accessible recipes and suggestions |
| Achieving broader impact | • Align with government and NGO-led initiatives such as national health guidelines and promoting campaigns e.g. Meat Free Monday, World Meat Free Day<br>• Influence demand for plant-based ingredients through buying practices via supplier requirements<br>• Promoting plant-based diets and cooking via school programmes<br>• Commissioning and promoting academic research that makes the case for plant-based diets<br>• Where supply is not available or sufficient, new partnerships and collaborations with retailers or manufacturers can be formed to develop new solutions at scale |

## International meat reduction support

The HSUS "Forward Food" initiative offers support and training to food service providers aiming to reduce meat consumption and increase plant-based meal options. The organisation and its international wing Humane Society International (HSI) support businesses wishing to integrate meat reduction into their sustainability plans, helping with monitoring and evaluation of the impact menu changes have on GHG emissions. The programme includes culinary training, advice on menus, marketing and product sourcing.

Kristie Middleton, Managing Director of HSUS's Farm Animal Protection programme, explains,

> There's no way we can achieve a humane society while continuing to consume meat, eggs, and dairy at our current levels. We're not just hoping for a growing demand for plant-based options: we're making the demand happen by reengineering menus to offer fewer animal-based foods and more phenomenally delicious plant-based options meant for the everyday consumer to enjoy.
>
> (Pers. Comm, 2018)

Forward Food is proving popular around the world and involves partnerships with influential businesses, including Compass Group and Aramark (Aramark, 2018; Compass Group, 2017). At the Extinction & Livestock conference Cheryl Queen, Compass Group's Vice President Communication and Corporate Affairs, explained how their chefs received training from HSUS to prepare veggie and vegan options that proved so successful they are now offered to guests in a range of sectors, including corporate, healthcare, colleges and 90% of universities supplied (Queen, 2017).

February 2018 saw HSI-Africa forming a partnership with one of Africa's largest universities, South Africa's University of Witwatersrand (WITS), which launched a new menu offering 23 plant-based dishes in its dining halls. One dining hall, the Green Hub, is the first dining hall in Africa to offer 100% plant-based dishes for breakfast, lunch and dinner every Monday. 2018 also saw HSI-Brazil announcing that four cities in North Eastern Brazil were committing to 100% plant-based cafeterias at public schools (HSI, 2018).

## Food for Life Served Here

Food for Life Served Here is an annually inspected award scheme for caterers that guarantees their menus meet standards of nutrition, sustainability and animal welfare. The award has three levels – bronze, silver and gold (Soil Association, 2018). Food for Life accredits over 1.8 million meals per day, over one million of which are at silver and gold levels, and certifies menus in over 50% of English primary schools, 30 universities and 50 hospitals as well as in business canteens, workplaces, and visitor attractions.

According to Strategy and Policy Director, Joanna Lewis, the vision for Food for Life Served Here is,

> to normalise healthy, sustainable and humane diets in all the places where people eat outside the home. Rewarding caterers for serving less but better meat is a central pillar of that, so alongside the standards promoting higher welfare we have standards at the Silver and Gold levels of the scheme promoting meat-free menus/days.
>
> (Pers. Comm, 2018)

It is very good to see that the scheme gives a heavy points weighting to meat-free menus and encourages support for meat-free themed days because of the opportunities presented to explain why it is important for health and the environment to eat less meat.

## Plant-based purchasing: a triple win for people, purse and planet

December 2017 saw Friends of the Earth US (FOE) and the RPN launching their municipal guide for climate-friendly purchasing showing how plant-based purchasing policies promote health, save money and benefit the environment (RPN and FOE, 2017). The report outlines practical actions that municipalities can take and includes advice on how to measure the climate and financial benefits of the policies as well as case studies of action already being taken.

Introducing the report Chloe Waterman, FOE's Senior Food Campaigner, cited case studies including that of Maricopa County, which saved $817,000 in a year by switching to more plant-based purchasing and four San Francisco hospitals, which saved $400,000 (RPN, 2017). Another case study outlines how, over a two-year period, a plant-based purchasing strategy saved one of California's largest school districts $42,000, reducing its food service carbon footprint by 14% and its water footprint by 6%. The carbon savings were the equivalent of covering all the roofs of its 85 schools with solar panels or driving 1.5 million miles a year. The water saving was 42 million gallons of water per school year (Hamerschlag and Kraus-Polk, 2017).

## Food policy model for the faiths

CreatureKind works to engage Christian institutions on the welfare of factory farmed animals. The organisation asks institutions to commit to a food sourcing policy that includes setting goals for progressively reducing the number of animal products served (CreatureKind, 2018). Founder David Clough, Professor of Theological Ethics at the University of Chester, advises that two organisations are currently signed up (University of Winchester and Friends House, London) and that he is currently in different stages of conversation with around 30 more, adding,

> There's a growing appreciation that Christians have faith-based reasons for caring about animal welfare, and that this demands a practical response in relation to the use we make of animals for food. We're finding many organisations who are keen to get help on where they can get started with strategies to reduce consumption of animal products and move to higher welfare sourcing for the remaining products.
>
> (Pers. Comm, 2018)

Friends House signed its partnership with CreatureKind in January 2018, committing to reduce the amount of meat, dairy and eggs served in its catering facilities by 20% within the next two years (Friends House, 2018).

## Innovative business award for reducing the number of animals in the supply chain

June 2018 saw Compassion in World Farming launching its Friendly Food Award aimed at reducing the number of animals and animal products purchased and produced by food companies by 25% by 2025 (Compassion in World Farming, 2018a). Positioning flexitarianism as a valuable business opportunity, the new corporate award will recognise three levels of commitment at bronze, silver and gold for a 10%, 15% or 25% reduction, respectively, in the number of animals purchased within five years.

Compassion hopes that this award will help the organisation achieve its strategic target of reducing the production and consumption of meat (including fish), milk and eggs by 50% in high-consuming nations by 2035 and by half globally (against 2015 baseline figures) by 2050 (Compassion in World Farming, 2018b).

Announcing the award Compassion's CEO Philip Lymbery commented,

> This ground-breaking award is hugely important in gaining recognition that the future wellbeing of people and animals, whether farmed or wildlife, really does hinge on reducing the number of farm animals being produced and in keeping them in better, more genuinely sustainable systems. We are currently using nearly half the useable land surface of the planet to grow our food; four-fifths of that land is producing meat and dairy, which contributes little more than a quarter of humanity's protein needs. By diversifying our protein sources, eating more plant proteins and less livestock products, and making sure that livestock products come from more nature-similar systems like pasture-fed, free range or organic, we can ensure that farm animal welfare is better served whilst saving the planet for future generations.
>
> (Lymbery, 2018)

## Call for a new global agreement on food and agriculture

The Extinction and Livestock conference saw Compassion unveiling its concept of a new global agreement to move to humane and sustainable food systems. The idea is aimed at breaking down policy silos and helping the international community to transform food systems to meet the objectives of existing Agreements in many areas, for example, the Paris Climate Change targets and those of the Convention on Biological Diversity.

Compassion conceives that a new global agreement would relate to both the production and consumption of food. Its objectives would include encouraging the adoption of healthy, sustainable and ethical diets; promoting the use of agricultural methods that improve the productivity and livelihoods of small-scale farmers; restoring and nurturing natural resources; reducing GHGs without having a detrimental impact on other objectives; minimising disease; minimising the use of antimicrobials, pesticides and herbicides;

enhancing food security including by reducing the use of human-edible crops as animal feed; enabling the restoration of wildlife habitats and respecting animal welfare (Compassion in World Farming, 2018c). The campaign for a global agreement is at the heart of the organisation's new strategic plan (Compassion in World Farming, 2018b).

## Conclusion and recommendations

Even if most governments are still lagging behind civil society and business action on sustainable and healthy diets, there are a plethora of exciting programmes and initiatives aimed at moving humankind away from the precipice. With further adoption and roll-out of the type of initiatives outlined in this chapter, along with those covered in other chapters, there is an abundance of hope for the future. And very importantly innovative development of alternative proteins is making them ever more delicious and affordable. In parallel the case for action to reduce production and consumption of animal products becomes ever more powerful with more and more organisations, institutions and businesses publishing research, reports, books and films raising awareness and driving change.

Moving forward it's imperative for future food and farming policies to be developed in a joined-up way involving, for example, health, environmental, animal welfare, conservation, consumer, producer and social justice bodies with action being founded on the introduction of sustainable dietary guidelines and an acknowledgement by all involved in the food system of the need for a move away from industrial livestock production towards farming systems producing humane and sustainable food. At the heart of any food policy should be the recognition that ending factory farming and eating more plants will benefit our health, the environment and animal welfare as well as securing a future for our children and grandchildren.

## References

Aramark, Aramark Launches New Plant-Based Culinary Training Curriculum with The Humane Society of the United States, 2018, accessed at: www.aramark.com/about-us/news/aramark-general/plant-based-culinary-training.

Barilla, 2015, accessed at: www.barillacfn.com/en/dissemination/milan_protocol/.

Barilla, 2016, accessed at: www.barillacfn.com/en/publications/second-edition-of-eating-planet/.

Barilla, 2017, accessed at: pyramid = www.barillacfn.com/en/dissemination/double_pyramid/.

Bojana Bajželj, Tim G. Benton, Michael Clark, Tara Garnett, Theresa M. Marteau, Keith S. Richards, Pete Smith & Milica Vasiljevic, 2015, Synergies between healthy and sustainable diets. Brief for GSDR 2015, accessed at: https://sustainabledevelopment.un.org/content/documents/635987-Bajzelj-Synergies%20between%20healthy%20and%20sustainable%20diets.pdf.

Brighton & Hove Food Partnership, 2012, Spade to spoon: Digging Deeper. A food strategy and action plan for Brighton & Hove.

Chatham House, 2017, China Shows Way with New Diet Guidelines on Meat, 21 June 2016, accessed at: www.chathamhouse.org/expert/comment/china-shows-way-new-diet-guidelines-meat.

Chile Nutritional Guidelines, 2013, accessed at: www.fao.org/nutrition/education/food-dietary-guidelines/regions/countries/chile/en/ and https://inta.cl/sites/default/files/_minisitios/consumidores/Revistas/guia_de_alimentacion.pd.

Chinese Nutrition Society, Chinese Dietary Guidelines, 2016, accessed at: http://dg.cnsoc.org/article/04/8a2389fd54b964c80154c1d781d90197.html and at FAO Dietary Guidelines site, www.fao.org/nutrition/education/food-dietary-guidelines/regions/countries/china/en/.

Cision P.R. Newswire, Pret Launches New Vegetarian Menu so Delicious it's Not Just For Veggies, 25 April 2017, accessed at: www.prnewswire.com/news-releases/pret-launches-new-vegetarian-menu-so-delicious-its-not-just-for-veggies-300444865.html.

City of Malmö, Policy for Sustainable Development and Food, 2010, accessed at: http://malmo.se/download/18.d8bc6b31373089f7d9800018573/Foodpolicy_Malmo.pdf.

Compass Group, Envision 2020, 2016, accessed at: https://issuu.com/becompass/docs/envision2020_zone_issuu?e=3949468/37377992.

Compass Group, Compass Group USA and Farm Animal Welfare, July 2017, accessed at: www.compass-group.com/content/dam/compass-group/corporate/Acting-responsibly/Our-pillars/CompassGroupUSAFarmAnimalWelfare2017_final.pdf.downloadasset.pdf.

Compass Group, Hospital Menus Add More Plant-Based Options to Meet Growing Demand, 2018, accessed at: www.compass-usa.com/school-hospital-menus-add-plant-based-options-meet-growing-demand/.

Compassion in World Farming, Friendly Food Award Criteria, 2018a, accessed at: www.friendlyfoodalliance.com.

Compassion in World Farming, Strategic Plan 2018–2022: Together, we are sparking a new era in food and farming, 2018b.

Compassion in World Farming, Why we need a UN Framework Convention on Food and Agriculture, 2018c.

Court, E. IKEA Vegan Hotdog Receives 95% Approval Rating, 14 May 2018, accessed at: www.plantbasednews.org/post/ikea-vegan-hotdog-receives-95-approval-rating.

CreatureKind, 2018 Institutional Food Policy, accessed at: www.becreaturekind.org/institutional-food-policy/.

De Vos, Jurriaan, M et al., Estimating the normal background rate of species extinction, *Conserv Biol* 2014; 29: 452–462.

Dibb, S. Eating Better for a Fair, Green and Healthier Future. Presentation at Extinction and Livestock Conference, 2017, accessed at: www.youtube.com/watch?v=I2qEt9Um-Ho&list=PL-7iZXkicZxfRMp9U7euR3GvhpZTR1V5y&t=0s&index=36.

Eating Better, The future of eating is flexitarian – companies leading the way, 2017.

Eating Better, Principles for eating meat and dairy more sustainably: the 'less and better' approach, 2018.

Emmott, S., 10 Billion, Penguin Books, 2013.

FAO, Final Document – International Scientific Symposium: Biodiversity and Sustainable Diets: 3–5 November 2010- Definition of Sustainable Diets. www.fao.org/ag/humannutrition/23781-0e8d8dc364ee46865d5841c48976e9980.pdf. Rome: Food and Agriculture Organisation, 2010.

Friends House, Friends House and CreatureKind Partner to Promote Compassionate Eating, 2018, accessed at: www.friendshouse.co.uk/news/friends-house-and-creaturekind-partner-promote-compassionate-eating.

Froggatt, A., and Wellesley, L., China Shows Way with New Diet Guidelines on Meat, Chatham House, 2016, accessed at: www.chathamhouse.org/expert/comment/china-shows-way-new-diet-guidelines-meat.

German Nutrition Society, 10 Guidelines of the German Nutrition Society, 2013, accessed at: www.fao.org/3/a-as683o.pdf.

Gonzalez Fischer, C., Garnett, T., Plates, pyramids, planet: Developments in national healthy and sustainable dietary guidelines: a state of play assessment, FAO and the Environmental Change Institute & The Oxford Martin Programme on the Future of Food, The University of Oxford, 2016.

Hamerschlag, K., and Kraus-Polk, J., Shrinking the Carbon and Water Footprint of School Food: A Recipe for Combating Climate Change. A pilot analysis of Oakland Unified School District's Food Programs, Friends of the Earth US, 2017.

Health Council of the Netherlands, Dutch Dietary Guidelines, 2015, accessed at: www.gezondheidsraad.nl/sites/default/files/201524edutch_dietary_guidelines_2015.pdf.

HSI, Four cities in Northeastern Brazil commit to adopting 100% plant-based cafeterias at public schools, 2018, accessed at: www.hsi.org/news/press_releases/2018/03/bahia-plant-based-schools-0321218.html.

HSI, Green Monday South Africa Institutional Culinary Training Program, 2018.

IKEA, People & Planet Positive: IKEA Group Sustainability Strategy for 2020, 2014, accessed at: www.ikea.com/gb/en/doc/general-document/ikea-download-our-sustainability-strategy-people-planet-positive-pdf__1364308374585.pdf.

Italian National Research Institute on Food and Nutrition, Guidelines for healthy Italian food habits, 2003, accessed at: www.fao.org/nutrition/education/food-based-dietary-guidelines/regions/countries/italy/en/.

Lang, T., Re-fashioning food systems with sustainable diet guidelines: towards a $SDG^2$ strategy, Centre for Food Policy, City, University of London, UK, March 2017, for Friends of the Earth, accessed at: http://foodresearch.org.uk/wp-content/uploads/2017/03/Sustainable_diets_January_2016_final.pdf.

Langin, K., 'Big Cats at a Tipping Point in the Wild, Jouberts Warn', Cat Watch, *National Geographic*, 7 Aug 2014. https://blog.nationalgeographic.org/2014/08/07/big-cats-at-a-tipping-point-in-the-wild-jouberts-warn/

Lymbery, P., *Dead Zone: Where the Wild Things Were*, Bloomsbury, 2017.

Lymbery, P., Speech at the Good Farm Animal Welfare Awards Ceremony, Paris, 22 June 2018.

Manchester Climate Change Agency, Manchester Climate Change Strategy 2017 – 50, 2016a.

Manchester Climate Change Agency, Manchester Climate Change Strategy Implementation Plan 2017 – 22, 2016b.

Malaysia, Nutritional Guidelines, 2010, accessed at: www.fao.org/nutrition/education/food-dietary-guidelines/regions/countries/malaysia/en/ and www.moh.gov.my/images/gallery/Garispanduan/diet/km6.pdf.

Mason, P., and Lang, T., *Sustainable Diets: How Ecological Nutrition Can Transform Consumption and the Food System*, Oxon: Routledge, 2017.

Meatless Monday, 2018a, accessed at: www.meatlessmonday.com/the-global-movement/.

Meatless Monday, 2018b, accessed at: www.meatlessmonday.com/meatless-monday-k-12/ and www.meatlessmonday.com/meatless-monday-campus/.

Meatless Monday, 2018c, accessed at: www.meatlessmonday.com/about-us/history/

Milan Urban Food Pact, 2015, accessed at: www.milanurbanfoodpolicypact.org.

Milman, O., and Leavenworth, S., "China's plan to cut meat consumption by 50% cheered by climate campaigners," The Guardian, 2016, accessed at: www.theguardian.com/world/2016/jun/20/chinas-meat-consumption-climate-change.

Ministry of Health, Eating Well with Canada's Food Guide, Canada, 2011, accessed at: www.hc-sc.gc.ca/fn-an/alt_formats/hpfb-dgpsa/pdf/food-guide-aliment/view_eatwell_vue_bienmang-eng.pdf.

Ministry of Health. *Eating and Activity Guidelines for New Zealand Adults*. 2015. Wellington: Ministry of Health, accessed at: www.health.govt.nz/system/files/documents/publications/eating-activity-guidelines-for-new-zealand-adults-oct15_0.pdf.

Ministry of Health. Mensajes y gráfica de las Guías Alimentarias para la Población Argentina, accessed at: www.msal.gob.ar/ent/index.php/component/content/article/9-informacion-ciudadanos/482-mensajes-y-grafica-de-las-guias-alimentarias-para-la-poblacion-argentina, 2015.

Ministry of Health of Brazil, Dietary Guidelines for the Brazilian Population, 2015, accessed at: http://bvsms.saude.gov.br/bvs/publicacoes/dietary_guidelines_brazilian_population.pdf.

Ministry of Health and Public Services, Israel, 2008, accessed at: www.fao.org/3/a-as685e.pdf.

Ministry of Health, Diet Guidelines, Uruguay, 2016, accessed at: www.msp.gub.uy/sites/default/files/archivos_adjuntos/MS_guia_web.pdf.

National Health and Medical Research Council, 2013 Australian Dietary Guidelines Summary. Canberra, accessed at: www.nhmrc.gov.au/_files_nhmrc/file/your_health/healthy/nutrition/n55a_australian_dietary_guidelines_summary_131014_1.pdf.

Netherlands Environmental Agency PBL, 2011, The Protein Puzzle, the Consumption and Production of Meat, Dairy and Fish in the EU, Annexes, www.pbl.nl/sites/default/files/cms/publicaties/Protein_Puzzle_Annexes.pdf.

Netherlands Nutrition Centre, Fact Sheet: Eating More Sustainably, 2017, accessed at: www.voedingscentrum.nl/Assets/Uploads/voedingscentrum/Documents/Professionals/Pers/Factsheets/English/Fact%20sheet_Eating%20more%20sustainably_2017.pdf.

Netherlands Nutrition Centre, Dietary Advice, 30 May 2018, accessed at: www.voedingscentrum.nl/nl/gezond-eten-met-de-schijf-van-vijf/wat-staat-er-in-de-vakken-van-de-schijf-van-vijf/vis-peulvruchten-vlees-ei-noten-en-zuivel.aspx.

OECD 2017, OECD-FAO Agricultural Outlook (Edition 2017) Retail weight per person per year, Table given at: https://data.oecd.org/agroutput/meat-consumption.htm.

PB2B, As Trend Turns into Seismic Shift PB2B presents its Top 5 Plant-based Predictions for 2018", 2018, accessed at: http://plantbased2b.wpengine.com/wp-content/uploads/2017/12/pb2b-trends.pdf.

Poore, J. and Nemecek, T., Reducing food's environmental impacts through producers and consumers, *Science* 360 (6393), 987 – 992. DOI: 10.1126/science.aaq0216

Public Health England, Press Release: New Eatwell Guide illustrates a healthy, balanced diet, 17 March 2016a.

Public Health England, The Eatwell Guide: How does it differ to the Eatwell Plate, and Why? March, 2016b.

Queen, C. Partnering for future food. Panel discussion at Extinction & Livestock Conference, CIWF, 2017, accessed at: www.youtube.com/watch?v=KAvBzJFu-kM&t=0s&list=PL-7iZXkicZxfRMp9U7euR3GvhpZTR1V5y&index=34.

Republic of Korea (South Korea), 2010, accessed at: www.fao.org/nutrition/education/food-dietary-guidelines/regions/countries/republic-of-korea/en/

Responsible Purchasing Network and Friends of the Earth US, The Meat of the Matter: A Municipal Guide to Climate – Friendly Food Purchasing, 2017, accessed at: https://1bps6437gg8c169i0y1drtgz-wpengine.netdna-ssl.com/wp-content/uploads/2017/12/MunicipalReport_ko_120117_v2-1.pdf.

Responsible Food Network, Webinar – The Meat of the Matter: A Municipal Guide to Climate-Friendly Food Purchasing Webinar Responsible Purchasing Network, 2017, accessed at: https://vimeo.com/247229207.

Searchinger, T. et al. 2013. “The Great Balancing Act.” Working Paper, Installment 1 of *Creating a Sustainable Food Future*. Washington, DC: World Resources Institute, accessed at: www.worldresourcesreport.org.

Soil Association, Standards for Food for Life Served Here, 2018, accessed at: www.soilassociation.org/certification/catering/business-support-for-award-holders/standards/.

Statista, 29 May 2018, accessed at: www.statista.com/statistics/679528/per-capita-meat-consumption-european-union-eu/

Sustainable Brands, Corporate Members, accessed at: www.sustainablebrands.com/members/corporate.

Sustainable Food Cities, The Sustainable Food Cities Award, 2018, accessed at: http://sustainablefoodcities.org/awards.

Turnwald, B.P., Boles, D.Z., Crum, A.J. Association between Indulgent Descriptions and Vegetable Consumption: Twisted Carrots and Dynamite Beets. *JAMA Intern Med*. 2017; 177(8): 1216–1218. doi:10.1001/jamainternmed.2017.1637.

United Nations, Department of Economic and Social Affairs, Population Division. The World’s Cities in 2016 – Data Booklet (ST/ESA/ SER.A/392), 2016.

U.S. Department of Health and Human Services (HHS) and the U.S. Department of Agriculture (USDA). 2015–2020 Dietary Guidelines for Americans. 2016, accessed at: https://health.gov/dietaryguidelines/2015/guidelines/chapter-1/key-recommendations/.

Vorster, H.H., Badham, J.B., Venter, C.S. An introduction to the revised food-based dietary guidelines for South Africa. *S Afr J Clin Nutr*. 2013; 26(3): S1–S164, accessed at: www.fao.org/3/a-as842e.pdf.

YouGov for Eating Better, 2017, accessed at: www.eating-better.org/blog/142/The-future-of-eating-is-flexitarian.html.

# Part VII

# Healthy and sustainable diets

# 27 Healthy and sustainable dietary patterns for prevention of chronic diseases and premature death

*Elena Hemler and Frank B. Hu*

Two of the most pressing threats to humanity are the exhaustion of our planet's resources and the increasing obesity epidemic and its related chronic diseases. Fortunately, dietary modifications can improve population health and sustainability, simultaneously addressing both problems. Healthy dietary patterns can reduce the risk of type 2 diabetes, obesity and cardiovascular disease (CVD) and typically have lower environmental impacts than unhealthy dietary patterns. However, global policy interventions are needed to make current food systems healthier and more sustainable to alleviate the burden of obesity and diet-related diseases (Malik, Willett, & Hu, 2013) as well as the environmental effects of food production (Dietary Guidelines Advisory Committee, 2015).

Since 1975, obesity rates have nearly tripled worldwide, driven by poor diets and decreasing physical activity levels (World Health Organization, 2017c). Globally, 39% of adults were overweight and 13% were obese in 2016 (World Health Organization, 2017c). Being overweight and/or obese puts individuals at greater risk for type 2 diabetes, CVD and other chronic diseases (World Health Organization, 2017c). Following the obesity trend, global diabetes rates have nearly doubled since 1980 (World Health Organization, 2017b). Diabetes is one of the leading causes of death worldwide, resulting in 1.6 million deaths in 2015 (World Health Organization, 2017d). CVD is the world's top killer (World Health Organization, 2017d), responsible for 31% of all deaths (World Health Organization, 2017a). Unlike obesity and diabetes, the prevalence of CVD has declined in some high-income countries over the past few decades (Roth et al., 2017). However, it has increased in several countries and has remained stable in most others, including low- and middle-income countries, where three-quarters of CVD deaths occur (Roth et al., 2017; World Health Organization, 2017a).

Poor diet quality and overconsumption of calories relative to energy expenditure are important factors driving the high burdens of obesity and chronic disease. After the Second World War, the US and other industrialised countries experienced major changes in the food system, which included a proliferation of highly processed foods loaded with unhealthy fats, added sugar and sodium (Popkin, Adair, & Ng, 2012). In recent decades, these changes have spread globally to many low- and middle- income countries (Mattei et al., 2015). Worldwide, traditional diets rich in vegetables, whole grains and other plant-based foods have transitioned to a Western diet, characterised by refined carbohydrates, added sugars, fat and animal-based foods (Popkin et al., 2012). As part of this nutrition transition, there have been major increases in animal products such as beef, pork, poultry and dairy in low- and middle-income countries (Delgado, 2003). Western-style dietary patterns (Fung et al., 2001; van Dam, Rimm, Willett, Stampfer, & Hu, 2002; Fung et al., 2004; Iqbal et al., 2008), as well as food typically included in this dietary pattern, such as sugar-sweetened beverages (Malik et al., 2010), fast food (Pan, Malik, & Hu, 2012) and red and processed meat (Fung et al., 2004), have been linked to negative health outcomes such as CVD and type 2 diabetes. While the specifics of the nutrition transition vary for each country, overall the world has moved towards unhealthy dietary patterns, resulting in increased burdens of obesity and chronic disease.

There is strong evidence that saturated fat (mainly found in animal foods) relative to unsaturated fats (mainly found in plant foods) is associated with higher risk of CVD (Li et al., 2015) and overall mortality (Wang et al., 2016). Plant foods, such as nuts, seeds and oils, are typically low in saturated fat and higher in monounsaturated and polyunsaturated fats. There is evidence that saturated fat is positively associated with a risk of premature death, while unsaturated fat, especially polyunsaturated fat, is associated with a decreased risk (Wang et al., 2016). However, when considering these associations, it is crucial to take into account what macronutrient or food is replacing saturated fat. In a study of 127,536 US women and men, replacing 5% of energy from saturated fats with equivalent energy from polyunsaturated fat, monounsaturated fat or whole grains was associated with a 25%, 15% and 9% lower risk of coronary heart disease, respectively (Li et al., 2015). Swapping saturated fat calories with the same amount of energy from refined starch and sugars did not change the risk of coronary heart disease (Li et al., 2015). These results suggest that exchanging animal-based foods high in saturated fats with healthy plant-based oils or foods can result in substantial health benefits, but replacing saturated fat with refined carbohydrates is not beneficial for CVD prevention.

There is also evidence that substituting animal protein with plant protein can reduce risk of chronic diseases and mortality (Pan et al., 2011; Pan, Sun, et al., 2012; Song et al., 2016). A study of US health professionals found that those who consumed more plant protein were less likely to die from any cause, while those who consumed more animal protein were more likely to die from cardiovascular causes (Song et al., 2016). These associations were stronger among participants who were overweight or obese, smoked, drank alcohol heavily or were physically inactive. In the same study, substituting eggs and processed or unprocessed red meat with plant protein reduced mortality risk (Song et al., 2016).

Additional evidence indicates that red meat consumption is associated with an increased risk of overall mortality as well as mortality from CVD and cancer (Pan, Sun, et al., 2012). Substituting one serving per day of red meat with other foods has been associated with a 7%–19% lower mortality risk, depending on the foods being substituted (Pan, Sun, et al., 2012). It is estimated that 9.3% of premature deaths in men and 7.6% in women could be prevented if everyone consumed less than half a serving of red meat per day (Pan, Sun, et al., 2012). Red meat, especially processed red meat, is also associated with type 2 diabetes (Pan et al., 2011). In one study, substituting one serving of red meat per day with one serving of nuts, low-fat dairy or whole grains was associated with 16%–35% lower risk of type 2 diabetes (Pan et al., 2011).

The 2015 Dietary Guidelines for Americans Advisory Committee reviewed the above studies and many others to inform recommendations for policymakers, nutrition educators and health professionals to prevent chronic diseases and promote health in the US. After thoroughly reviewing scientific evidence, modelling the effects of potential dietary changes and analysing existing data, the committee recommended focussing on a healthy overall dietary pattern, rather than individual food groups or nutrients (Dietary Guidelines Advisory Committee, 2015). This dietary pattern can be achieved in many different ways and should be customised based on individual food preferences and health conditions. The committee also recommended limiting saturated fat to less than 10% of calories and replacing it with unsaturated fat, particularly polyunsaturated fat. They identified that dietary patterns associated with beneficial health outcomes commonly included higher intake of vegetables, fruits, whole grains, low- or non-fat dairy, seafood, legumes and nuts. Healthy dietary patterns also included lower consumption of red and processed meat, refined grains and sugar-sweetened foods and beverages (Dietary Guidelines Advisory Committee, 2015).

Dietary patterns that contain higher amounts of plant-based foods tend to be better for us, and they are also better for our planet (Dietary Guidelines Advisory Committee, 2015; Nelson, Hamm, Hu, Abrams, & Griffin, 2016). Food production is the largest contributor to the planet's loss of biodiversity and is also responsible for over 70% of fresh water use, 80% of deforestation and up to 30% of greenhouse gas emissions from humans (Nelson et al., 2016). Numerous studies have shown that diets high in animal-based foods have more environmental impact than those high in plant-based foods (Nelson et al., 2016). Producing animal products is energy-intensive; one calorie of beef requires 40 calories of fossil fuel, one calorie of milk requires 14, while one calorie of grains only requires 2.2 (Dietary Guidelines Advisory Committee, 2015). Although many animal products have a larger environmental impact than plant products, meat has the largest impact (Nelson et al., 2016). Multiple studies have indicated that reducing meat consumption could conserve land, energy and water and decrease greenhouse gas emissions, as well as improve health (Nelson et al., 2016). One study examined actual diets in the UK and found that greenhouse gas emissions per individual decreased as meat consumption decreased (Scarborough et al., 2014). Participants who ate meat frequently had associated greenhouse emissions that were twice as high as vegans (Scarborough et al., 2014).

However, it is not necessary to eliminate animal products to achieve a sustainable diet. Dietary patterns that follow the US Dietary Guidelines, such as the Mediterranean and Dietary Approaches to Stop Hypertension patterns, include animal products and allow small amounts of red meat. Both of these dietary patterns are more sustainable than average US diets and their beneficial health effects have been established (Nelson et al., 2016). It is important to note, a dietary pattern that replaces animal products with refined grains or added sugar can be sustainable, but will not confer the same health benefits as patterns that substitute vegetables, whole grains, nuts and legumes for animal products. Healthy plant-based diets have been associated with a decreased risk of developing coronary heart disease, while unhealthy plant-based diets high in refined grains, added sugars and potatoes have been associated with an increased risk (Satija et al., 2017). Additionally, very high-calorie diets, which are common in high-income countries, not only cause obesity but are also more environmentally damaging than lower energy diets (Nelson et al., 2016). In countries such as the US, the population should reduce total energy consumption to achieve a diet that is more sustainable and healthier (Dietary Guidelines Advisory Committee, 2015). Altogether, the evidence shows that healthy dietary patterns that include plant-based foods and limit animal-based foods can benefit population health and our environment.

As the world's population continues to grow and climate change progresses, our global resources will be increasingly strained, threatening food security (Nelson et al., 2016). The epidemics of obesity and diet-related chronic diseases are additional public health risks, and will continue to worsen if no action is taken (Kelly, Yang, Chen, Reynolds, & He, 2008; Sacco et al., 2016; International Diabetes Federation, 2017). Shifting populations to healthier, plant-based dietary patterns can alleviate these threats. To achieve global change, large-scale interventions are necessary to enable and incentivise individuals to consume high-quality diets, reduce energy intake and minimise waste (Dietary Guidelines Advisory Committee, 2015). Policy changes are also crucial to support sustainable food production and create a food environment that encourages healthy and sustainable choices (Dietary Guidelines Advisory Committee, 2015). Although the benefits of shifting dietary patterns are evident, much work is needed to translate the evidence into policies and interventions to mitigate the growing diet-related epidemics and future environmental crises (Figure 27.1).

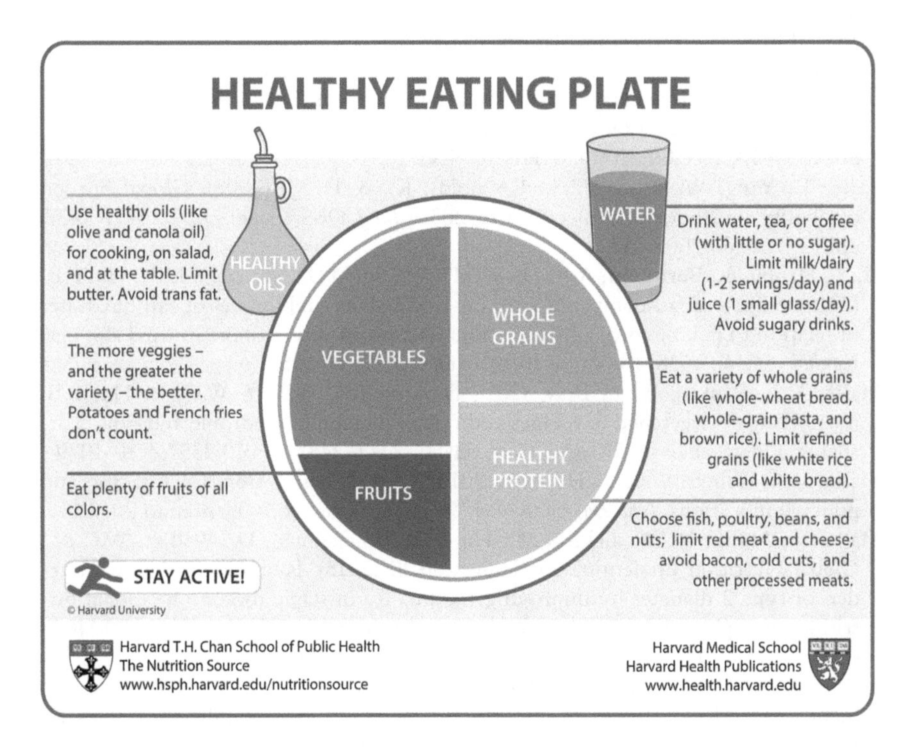


*Figure 27.1* The healthy eating plate.



## References

Delgado, C.L. (2003). Rising consumption of meat and milk in developing countries has created a new food revolution. *J Nutr, 133*(11 Suppl 2), 3907S–3910S.

Dietary Guidelines Advisory Committee. (2015). *Scientific Report of the 2015 Dietary Guidelines Advisory Committee.* Retrieved from Washington, DC. Retrieved from: https://health.gov/dietaryguidelines/2015-scientific-report/PDFs/Scientific-Report-of-the-2015-Dietary-Guidelines-Advisory-Committee.pdf.

Fung, T.T., Rimm, E.B., Spiegelman, D., Rifai, N., Tofler, G.H., Willett, W.C., & Hu, F.B. (2001). Association between dietary patterns and plasma biomarkers of obesity and cardiovascular disease risk. *Am J Clin Nutr, 73*(1), 61–67.

Fung, T.T., Schulze, M., Manson, J.E., Willett, W.C., & Hu, F.B. (2004). Dietary patterns, meat intake, and the risk of type 2 diabetes in women. *Arch Intern Med, 164*(20), 2235–2240. doi:10.1001/archinte.164.20.2235.

International Diabetes Federation. (2017). *IDF Diabetes Atlas Eighth Edition*. Retrieved from: www.diabetesatlas.org/

Iqbal, R., Anand, S., Ounpuu, S., Islam, S., Zhang, X., Rangarajan, S., ... Investigators, I.S. (2008). Dietary patterns and the risk of acute myocardial infarction in 52 countries: results of the INTERHEART study. *Circulation, 118*(19), 1929–1937. doi:10.1161/CIRCULATIONAHA.107.738716.

Kelly, T., Yang, W., Chen, C.S., Reynolds, K., & He, J. (2008). Global burden of obesity in 2005 and projections to 2030. *Int J Obes (Lond), 32*(9), 1431–1437. doi:10.1038/ijo.2008.102.

Li, Y., Hruby, A., Bernstein, A.M., Ley, S.H., Wang, D.D., Chiuve, S.E., ... Hu, F.B. (2015). Saturated fats compared with unsaturated fats and sources of carbohydrates in relation to risk of coronary heart disease: a prospective cohort study. *J Am Coll Cardiol, 66*(14), 1538–1548. doi:10.1016/j.jacc.2015.07.055.

Malik, V.S., Popkin, B.M., Bray, G.A., Despres, J.P., Willett, W.C., & Hu, F.B. (2010). Sugar-sweetened beverages and risk of metabolic syndrome and type 2 diabetes: a meta-analysis. *Diabetes Care, 33*(11), 2477–2483. doi:10.2337/dc10-1079.

Malik, V.S., Willett, W.C., & Hu, F.B. (2013). Global obesity: trends, risk factors and policy implications. *Nat Rev Endocrinol, 9*(1), 13–27. doi:10.1038/nrendo.2012.199.

Mattei, J., Malik, V., Wedick, N.M., Hu, F.B., Spiegelman, D., Willett, W.C., ... Global Nutrition Epidemiologic Transition, I. (2015). Reducing the global burden of type 2 diabetes by improving the quality of staple foods: The global nutrition and epidemiologic transition initiative. *Global Health, 11*, 23. doi:10.1186/s12992-015-0109-9.

Nelson, M.E., Hamm, M.W., Hu, F.B., Abrams, S.A., & Griffin, T.S. (2016). Alignment of healthy dietary patterns and environmental sustainability: a systematic review. *Adv Nutr,* 7(6), 1005–1025. doi:10.3945/an.116.012567.

Pan, A., Sun, Q., Bernstein, A.M., Schulze, M.B., Manson, J.E., Willett, W.C., & Hu, F.B. (2011). Red meat consumption and risk of type 2 diabetes: 3 cohorts of US adults and an updated meta-analysis. *Am J Clin Nutr, 94*(4), 1088–1096. doi:10.3945/ajcn.111.018978.

Pan, A., Malik, V.S., & Hu, F.B. (2012). Exporting diabetes mellitus to Asia: the impact of Western-style fast food. *Circulation, 126*(2), 163–165. doi:10.1161/CIRCULATIONAHA.112.115923.

Pan, A., Sun, Q., Bernstein, A.M., Schulze, M.B., Manson, J.E., Stampfer, M.J., ... Hu, F.B. (2012). Red meat consumption and mortality: results from 2 prospective cohort studies. *Arch Intern Med, 172*(7), 555–563. doi:10.1001/archinternmed.2011.2287.

Popkin, B.M., Adair, L.S., & Ng, S.W. (2012). Global nutrition transition and the pandemic of obesity in developing countries. *Nutr Rev, 70*(1), 3–21. doi:10.1111/j.1753–4887.2011.00456.x.

Roth, G.A., Johnson, C., Abajobir, A., Abd-Allah, F., Abera, S.F., Abyu, G., ... Murray, C. (2017). Global, regional, and national burden of cardiovascular diseases for 10 causes, 1990 to 2015. *J Am Coll Cardiol, 70*(1), 1–25. doi:10.1016/j.jacc.2017.04.052.

Sacco, R.L., Roth, G.A., Reddy, K.S., Arnett, D.K., Bonita, R., Gaziano, T.A., ... Zoghbi, W.A. (2016). The heart of 25 by 25: achieving the goal of reducing global and regional premature deaths from cardiovascular diseases and stroke: a modeling study from the American Heart Association and World Heart Federation. *Glob Heart, 11*(2), 251–264. doi:10.1016/j.gheart.2016.04.002.

Satija, A., Bhupathiraju, S.N., Spiegelman, D., Chiuve, S.E., Manson, J.E., Willett, W., ... Hu, F.B. (2017). Healthful and unhealthful plant-based diets and the risk of coronary heart disease in U.S. adults. *J Am Coll Cardiol, 70*(4), 411–422. doi:10.1016/j.jacc.2017.05.047.

Scarborough, P., Appleby, P.N., Mizdrak, A., Briggs, A.D., Travis, R.C., Bradbury, K.E., & Key, T.J. (2014). Dietary greenhouse gas emissions of meat-eaters, fish-eaters, vegetarians and vegans in the UK. *Clim Change, 125*(2), 179–192. doi:10.1007/s10584-014-1169-1.

Song, M., Fung, T.T., Hu, F.B., Willett, W.C., Longo, V.D., Chan, A.T., & Giovannucci, E.L. (2016). Association of animal and plant protein intake with all-cause and cause-specific mortality. *JAMA Intern Med, 176*(10), 1453–1463. doi:10.1001/jamainternmed.2016.4182.

van Dam, R.M., Rimm, E.B., Willett, W.C., Stampfer, M.J., & Hu, F.B. (2002). Dietary patterns and risk for type 2 diabetes mellitus in U.S. men. *Ann Intern Med, 136*(3), 201–209.

Wang, D.D., Li, Y., Chiuve, S.E., Stampfer, M.J., Manson, J.E., Rimm, E.B., ... Hu, F.B. (2016). Association of specific dietary fats with total and cause-specific mortality. *JAMA Intern Med, 176*(8), 1134–1145. doi:10.1001/jamainternmed.2016.2417.

World Health Organization. (2017a). Cardiovascular diseases. Retrieved from: www.who.int/cardiovascular_diseases/en/.

World Health Organization. (2017b, November). Diabetes. Retrieved from: www.who.int/mediacentre/factsheets/fs312/en/.

World Health Organization. (2017c, October). Obesity and overweight fact sheet. Retrieved from: www.who.int/mediacentre/factsheets/fs311/en/.

World Health Organization. (2017d, January 2017). The top 10 causes of death. Retrieved from: www.who.int/mediacentre/factsheets/fs310/en/.

# 28 The threat to public health from livestock production

*Aysha Akhtar*

As a neurologist and public health specialist, I've noticed a tendency in medicine – and in the world at large – to pit humans against animals. There is a misconception that we can use our resources and time either to protect humans or to protect animals. In other words, we have been led to believe we have to make a choice: it's us or the animals.

Throughout my career though, I have found that it's quite the opposite: our fates are shared with those of other animals. And nowhere is this more apparent than in our eating habits.

Humans are eating more animals than ever before. Today, more than 74 billion land animals are raised and killed for food worldwide annually. In the US alone, more than nine billion land animals are slaughtered every year for food – that's about one million per hour!

With this demand for animal products, comes the intensification of animal agriculture and the suffering of more animals than ever before seen in human history. The animal agriculture industry has chosen to sacrifice the space and well-being of animals in the name of efficiency. Farmed animals are treated as "production units" and are denied their most basic needs. The overwhelming majority of animals raised for food are housed in overcrowded conditions without access to fresh air, sunlight or room to move about normally (see Joyce D'Silva's Chapter 11 for more details). This demand-driven transformation of animal agriculture is so dramatic that it has been dubbed the "Livestock Revolution".

The conditions in these farms greatly contribute to the creation of deadly pathogens, including influenza viruses. Here's how it works: wild aquatic birds are the primordial source of all influenza A viruses – the ones that have the potential to cause pandemics. However people rarely become infected directly from aquatic birds. Usually, an intermediate host must be involved – providing the right biological setting for the virus to transform into something that can easily infect humans. And that's where chickens and other farmed animals come in.

Most avian influenzas are mild viruses and not very dangerous. However, once they enter poultry factory farms, they can rapidly mutate into very lethal forms. Viruses pass readily from animal to animal in these operations not only because the animals are crowded together but also because their ability to fight off infections is severely reduced by the distressing situations in which they are forced to live. In another twist, pigs are commonly referred to as "mixing vessels" in whom influenza viruses from birds, humans and other pigs co-mingle, leading to another way the virus can mutate. Every transmission of the virus from one animal to another brings us closer to a pandemic.

New and dangerous pathogens are emerging and spreading rapidly throughout the world at an ever-increasing pace. Novel influenza A viruses such as H5N1 and H7N9 have already caused a significant number of human illnesses and deaths. So far, we have been lucky, however, because although many of these influenza strains are very lethal, they have not been very contagious among humans.

In spite of our current low risk, it is just a matter of time before an influenza strain evolves into a contagious form that leads to a pandemic. As long as we continue to confine billions of animals on factory farms, we are unwittingly creating a global laboratory for the evolution of the virus.

If we really want to prevent a deadly virus from causing a pandemic, we need to turn our attention to how we treat other animals and to our global appetite for animal products. We don't need to look for complicated strategies that at best are short term and narrowly address the global problem we face. The only lasting solution is to significantly reduce our eating of animals. By becoming a vegetarian, or better yet a vegan, each one of us can decrease the number of animals crowded into factory farms and help prevent the emergence of dangerous pathogens.

This strategy has even farther-reaching implications. Our meat addiction is literally killing us. As explained by Elena Hemler and Dr Frank Hu in Chapter 27, overconsumption of meat is now known to be a major player in the deadly rise in obesity and chronic diseases such as heart disease and stroke, and in environmental destruction and climate change.

By simply choosing one plate of food over another, we can each single-handedly help thwart a pandemic, reduce climate change, keep ourselves healthier and reduce animal suffering. I challenge anyone to name a drug that can accomplish all of that!

We have traditionally viewed our health in an evolutionary vacuum – ignoring its connection with our relationships with other animals. But this needs to change. Our health is inextricably linked with how we treat animals. In other words, what's good for animals is also what's good for us.

The opinions expressed here are solely those of Dr Akhtar.

# 29 Food, feed and sustainable diets

*Duncan Williamson*

Biodiversity is disappearing at an astonishing rate because of the food we eat; however biodiversity is key to a resilient food system. In a world where more and more people adopt a Western diet – one that's high in meat, dairy and ultra-processed food – livestock and the crops and pastures to feed them will continue to put an enormous strain on our natural resources. Current pressures on the planet will further increase with population and economic growth, unless consumption and production patterns become more efficient, less polluting and are brought to operate within planetary boundaries. They must also deliver healthy, culturally diverse, delicious dishes. Food systems rely on a natural resource base that is extremely vulnerable to climate change and biodiversity loss, both of which are undermining food security and nutrition.

WWF-UK is looking at the impacts animal protein has on our planet, in particular the often-hidden impacts of animal feed. As part of this process we published Appetite for Destruction (Halevy & Ukropcova, 2017). This is based on two reports prepared for WWF in 2016: "A risk benefit analysis of mariculture" by ABP Marine Environmental Research and "The Environmental impacts of livestock feed" by 3Keel LLP. These reports took us on a journey from farming the oceans to land-use change, through soy to our plates. They showed us that not only is business as usual not desirable, it won't be possible from a climate, water and land perspective. We'll need to go beyond production improvements and waste reduction and innovate and look at the food on our plates.

We already produce enough to feed the world, but overconsumption, inequality, waste and inadequate production and distribution systems stand in the way of enough food for everyone and space for wildlife. To feed the world within our planetary boundaries, we need to consume and produce food differently. The systems in which we rear animals require inputs such as feed and fertiliser. As we eat more animal products, we also need more animal feed. Protein-rich soy is such an important feed ingredient that the average European consumes approximately 61 kg of soy per year, largely indirectly through the animal products that they eat like chicken, pork, salmon, cheese, milk and eggs. Feed production eats up land: in 2010 the British livestock-related soy consumption alone required an area almost the size of Yorkshire.

If the global demand for animal products grows as anticipated, soy production would need to increase by nearly 80% to 390 million tonnes to feed all the animals destined for our plates. Increased feed production requires land but it's not clear where this land can be found. Today agricultural land occupies about 38% of Earth's terrestrial surface, and converting or changing land uses could have severe impacts on ecosystems, the biodiversity and the services they provide, and threaten food and water security for local communities.

Producing more on the same land is also a daunting challenge as the collective effect of intensification has meant increased environmental impacts. These impacts are more pronounced if intensification efforts concentrate only on producing more on less land by increasing other inputs, such as fertilisers or water. Intensive livestock production relies on chemical fertilisers, pesticides, unsustainable water abstraction and preventive use of antibiotics. It often leads to negative outcomes and vulnerabilities, which threaten the ecosystem services we depend on.

Feed crops are already produced in a large number of Earth's most valuable and vulnerable areas, such as the Amazon, Cerrado, Congo Basin, Yangtze, Mekong, Himalayas and the Deccan Plateau forests. Fifty years ago, the Cerrado was probably the most biodiverse grassland on the planet. This magnificent grassland was home to jaguars, armadillos, giant anteaters, Spix macaws and tapir, who shared with cattle. It had at least 5,000 endemic plant species. Since then, three-quarters of the Cerrado has disappeared under the plough. Today it produces 70% of Brazil's crops. Brazil is the world's biggest exporter of soy, beef, cotton, coffee, chickens, sugar, ethanol, tobacco and orange juice. We have no way of knowing how many endemic plant species have disappeared so we can grow soy.

## Who eats all this soy?

Globally, the biggest user of crop-based feed, with 41.5% of the world's feed in 2009, is the poultry industry. Poultry increased its share of world meat production from 15% in the mid-1960s to 32% by 2012 as per capita consumption increased threefold. In 2014, there were over 23 billion chickens, turkeys, geese, ducks and guinea fowl on the planet at any one time – more than three per person.

The second largest feed crop consumer, with 30% of the world's feed in 2009, is the pig industry. In the UK, pork is the second favourite meat after chicken, with each person eating on average 25 kg a year in 2015. That is nearly the whole recommended yearly intake for all meats. With the rapid expansion of the middle class in China over half the world's pork is produced and consumed in China (Halevy & Ukropcova, 2017).

Capture fisheries and aquaculture provide three billion people with almost 20% of their average per capita intake of animal protein and a further 1.3 billion people with about 15% of their per capita intake. Aquaculture used 4% of the world's feed in 2009. Global consumption is on the rise and is gaining more traction due to its nutritional value and health benefits. The global average consumption of fish per person per year has almost doubled in the last 50 years, from 9.9 kg in the 1960s to 19.7 kg in 2013. Meeting projections for seafood demand in 2050 would require a 200% increase compared to 2012 levels or 271 million tonnes. With wild fish stocks already under immense pressure, this increased demand, if realised, would have to be met through aquaculture. Aquaculture has seen a shift towards industrialisation. In 2012, 70% of total aquaculture production was based on feed supplements, including soy and maize. To grow these feeds, aquaculture indirectly used an additional 26.4 million hectares in 2010 – an area about the size of the UK (Halevy & Ukropcova, 2017).

## Sustainable diets

Significant environmental benefits could be achieved by simply sticking to the nutritionally recommended amount of protein. If everyone reduced the amount of animal products that they ate to suit their nutritional requirements, the total agricultural land use would decline by 13%. That means nearly 650 million hectares – or an area nearly 27 times bigger than the UK – would be saved from agricultural production.

WWF's Livewell Plates and six Livewell principles (Eat more plants, eat a variety of foods, waste less, moderate your meat consumption, buy food from a credible certified source and eat less foods high in salt, sugar and fat) are good guides to how to put this into action (WWF, 2017). They illustrate how we can ensure people get their necessary nutritional requirements without a further increase in agricultural land area and still contribute towards meeting the Paris Agreement commitment to keep global warming well below 2°C. Eating less animal protein would also make production systems with lower environmental impact and healthier, more nutritious outputs much more possible.

## New sources of feed

WWF are at the forefront of feed innovation. Project-X's FEED-X is a WWF programme which aims to help the aquaculture industry identify, test and scale-up new alternative feed ingredients, including insects and algae. It'll seek to drive consensus on the most scalable and sustainable feeds and accelerate the uptake of these ingredients.

Algae can grow in a range of aquatic environments and come in either microform or macroform. Microalgae are microscopic, often independent single-celled organisms. Macroalgae or seaweeds are organisms which absorb nutrients from the surrounding aquatic environment through their whole surface. Algal growth is relatively straightforward, as they only need a basic form of energy (such as light and sugars), $CO_2$, water and a few inorganic nutrients to grow.

In many areas of the world insects are considered a delicacy. But flies, crickets and other insects could also reduce the mounting pressure on land and biodiversity through use as feed or food. Insects have a lower environmental impact than soy feed and animal products, producing the same amount of edible protein with less land, lower greenhouse gas emissions and similar amounts of energy. Insects as feed show great promise too. When insects have been used to supplement livestock feed, feeding trials have revealed that piglets' gut health was improved, while chickens fed on insect protein (which is part of their natural diet) performed as well as those given current commercial feeds; research also shows that insect meal could replace up to 50% of fish feed without affecting animal performance.

## Conclusion

We know ecosystems are threatened by the current food system. We often hear about the amount of land in Africa available for agriculture. Many African governments are eyeing the agriculture opportunities available by following the South American model. Africa has a greater proportion of grasslands intact than any other continent. The Serengeti of northern Tanzania is home to the greatest concentration of large mammals in the world. Buffalo, rhino, elephants, zebras and giraffes graze there. Around 1.2 million wildebeest follow the rains that maintain the grasses. Their migration, crossing crocodile-infested rivers and braving prides of hunting lions, is one of the planet's great wildlife spectacles. This might disappear if our appetite for ever more animal production and consumption takes hold in Africa.

It's possible to have a food system that provides us with nutritionally and environmentally sustainable food whilst leaving room for nature to thrive. Adopting a healthy, sustainable diet is one part of the solution and we believe our six Livewell principles form the basis of a sustainable diet and a well-functioning food system. We need to produce food and feed differently. To achieve this vision, we must move out of silos and collaborate cross sartorially with business and policymakers to create a food system that provides us with healthy food, sustainable feed and thriving biodiversity.

## References

Eating for 2 degrees – New and Updated Livewell plates, WWF Summary Report, August 2017, accessed at: www.wwf.org.uk/eatingfor2degrees.

Halevy, S., & Ukropcova, V. (2017) *Appetite for Destruction*, accessed at: www.wwf.org.uk/updates/appetite-for-destruction.

Report for WWF UK - *A risk benefit analysis of mariculture as a means to reduce the impacts of terrestrial production of food and energy* by ABP Marine Environmental Research Ltd, accessed at: www.sarf.org.uk/cms-assets/documents/232492-618987.sarf106.pdf.

# 30 Conserving biodiversity through sustainable diets

*Glyn Davies*

Looking at humanity's footprint on the world, the pressures and threats from human consumption are all rising rapidly, including world population, freshwater consumption, fertiliser use and carbon dioxide emissions. An important driver of these statistics is the food system, which is currently unsustainable: improvements must be sought at both ends of the supply chain, to achieve both sustainable consumption and sustainable production.

An important part of the World Wildlife Fund (WWF)'s approach to addressing these issues has been through engagement with businesses. We consider businesses as important agents of change to achieve a world where people and nature thrive and have worked with them for decades to advance a sustainability agenda. This has included developing standards of sustainable production for timber (FSC), seafood (MSC), palm oil (RSPO) and soy (RTRS). These approaches to "certified production" have had many successes, yet also have limitations that need to be addressed via government regulation and consumer behaviour change.

WWF has developed a strong narrative on sustainable consumption through the Livewell Diets Principles (WWF, 2017). These emphasise enjoying vegetables and whole grains, having a colourful and varied plate, producing less waste and substantially moderating our meat consumption. We encourage purchase of credibly certified foods and support a diet with fewer items high in fat, salt and sugar. Through our partnership with Sodexo we have been working to reduce meat and increase vegetables in their Green & Lean meals programme, with successful uptake in a pilot programme with eight schools in 2015 (Sodexo, 2015). During that time 20,000 school meals had almost a ton of meat replaced by extra vegetables, and this was well received by the students!

At the other end of the food chain, a number of efforts have been made to support sustainable production. There are a number of successful examples of regulation and voluntary certification, leading to recovery of fish stocks – such as the Alaskan salmon and North Sea cod fisheries, both of which are MSC certified with the latter being an important part of WWF's partnership with Marks & Spencer (M&S) and their Plan A. As increasing amounts of fish consumed comes from aquaculture, more fish farms need to be Aquaculture

Stewardship Certified (ASC; currently only 1.5% of aquaculture product is ASC certified), paying particular attention to fish feed.

Certification of agricultural crops, for example soy (RTRS) and palm oil (RSPO), has had substantial effects in supporting companies to be more sustainable but has had limited impact in reducing forest and biodiversity loss. These approaches to sustainable production will only be effective if they are linked to spatial plans for conservation and sustainable development, and supported by stronger regulation. For example, the Soy Moratorium in the Amazon has substantially reduced deforestation in that biome but has resulted in increased habitat loss in the adjacent Cerrado region. The recent Cerrado Manifesto Commitment by companies and non-governmental organisations (NGOs) in Brazil has addressed this problem of "leakage" and is therefore bringing conservation benefits to the Cerrado region, as explained by WWF's Jean Francois Timmers in Chapter 7.

Whilst these certification tools are available, communication on sustainable diets is important to support conservation on our planet. And one way to engage with people who don't like being told what to eat is to explain the health benefits for individuals and communities of the Livewell Diet. They then understand the double benefits of healthy people and a healthy planet.

## References

Sodexo Press Release, Sodexo and WWF pilot Green and Lean meals Sodexo, November 18 2015, accessed at: https://uk.sodexo.com/home/media/press-releases/newsListArea/uk-press-releases/sodexo-and-wwf-pilot-green-and-l.html.

WWF Summary Report, Eating for 2 Degrees, New and Updated Livewell Plates, August 2017, accessed at: www.wwf.org.uk/what-we-do/area-of-work/livewell.

# 31 The case for veganism

*Andrew Knight and Jasmijn de Boo*

## Respect for life

The multiplicity of species of animals with which we share the Earth exhibit a truly remarkable array of characteristics and biological adaptations. Many of them are sentient, have complex emotional and social lives and exist within ecological webs of intricate complexity. Animals such as primates, cetaceans and corvids have demonstrated surprising linguistic and other communicative abilities; exhibit complex, socially transmitted behaviour; and have advanced cognitive capacities (Benz-Schwarzburg and Knight 2011).

Whichever characteristics we might reasonably consider necessary for justifying moral consideration, it seems that some animals possess them, at least to some morally significant degree. The more we learn from studies in ethology, cognition and related fields, the more it appears that the differences between us and many other animal species are merely differences of degree, rather than fundamental differences of kind.

All animals have an interest in continuing to live and in avoiding harm, danger and death, regardless of cognitive abilities. The uncomfortable truth is that animal farming and killing causes widespread animal suffering and that this occurs primarily to satisfy human dietary preferences, rather than to fulfil essential needs. The annual farming of over 70 billion terrestrial animals globally (FAO 2017), and up to three trillion fish and other marine animals (Anon 2014), violates their moral rights to exercise their own preferences, in pursuit of their own interests, and indeed – when they are killed – to live at all.

Why, then, is there not wider recognition of the need for human dietary change? Vegan diets are solely plant-based, eschewing animal products such as meat, milk and eggs. They have a remarkable ability to concurrently address many of the most serious problems presently facing humanity and the other species at the mercy of our lifestyle choices.

## Carnism and social justice

One of the main reasons why veganism – although rapidly growing – is not yet mainstream, is because of carnism. Joy (2018) has defined this as the underlying belief system, or ideology, that conditions people to eat certain animals, such as cows, pigs and chickens, but not others, such as cats and dogs. According to Joy (2018),

> people rarely realize that eating animals is a choice, rather than a given. In meat-eating cultures around the world, people typically don't think about why they eat certain animals but not others, or why they eat any animals at all. But when eating animals is not a necessity, which is the case for many people in the world today, then it is a choice – and choices always stem from beliefs. [...] Carnism is structured like other systems of oppression, such as racism, sexism, and heterosexism. While the experience of each set of victims of oppressive systems will always be unique, the systems are similar because *the mentality that enables the oppression is the same.*
>
> Ultimately, cultivating compassion and justice is not simply about changing behaviors; it is about changing consciousness so that no "others", human or nonhuman, are victims of oppression. To bring about a more compassionate and just society, then, we must strive to include all forms of oppression in our awareness, including carnism.

The systemic exploitation of non-human animals also entraps many humans: for example those who work in slaughterhouses or in oppressive conditions in the fishing industries of certain nations.

## Global public health

Food consumption is not only a basic activity necessary to sustain individual life but also an important cultural activity. Preparing and sharing food are social events. However, the nature of food production and consumption has changed significantly over the past 40 years: from traditional family meals to food on the go, from slow to fast food and from a largely unrefined complex carbohydrate diet full of fibre, micronutrients, sufficient protein and small amounts of fat and sugar, to one in which foodstuffs have been stripped of their nutritional content and processed into calorie-dense and nutrient-poor items. The addition of unhealthy fats, salt, sugar and additives, and the replacement of plant proteins with animal proteins have turned many food and drink items into serious risk factors for human health.

According to the World Health Organisation (WHO) (2017a), the number one cause of death worldwide is ischaemic heart disease and stroke, accounting for a combined 15 million deaths in 2015.

As early as the 1960s, the link between high saturated fat and blood cholesterol and the prevalence of coronary heart disease (CHD) became apparent. In 1958, Dr Ancel Keys launched a comparative study in seven countries to study the effect of diet on CHD. Dr Keys and colleagues demonstrated the health protective effects of the "Mediterranean" diet, that is, one high in fruit and vegetable intake, grains and legumes, with some olive oil and nuts, and some small amounts of fish and alcohol. Later studies confirmed that the main benefits to reducing the risk of CHD were derived from the plant-based components of the diet rather than the olive oil or fish (e.g. as summarised in Esselstyn 2017).

Another global epidemic is obesity, which frequently causes secondary health problems, such as type 2 diabetes and hypertension. According to the WHO (2017b) more than 1.9 billion adults, 18 years and older, were overweight in 2016. Of these over 650 million were obese. In all, 41 million children under the age of 5 were overweight or obese in 2016, and over 340 million children and adolescents aged 5–19 were overweight or obese in 2016. Obesity is preventable. Nearly a billion people worldwide are malnourished or underweight.

Additionally, studies consistently demonstrate that those eating plant-strong diets generally have:

- Lower blood pressure and lower incidence of cardiovascular disease (e.g. Leenders et al. 2013).
- Lower cholesterol levels (e.g. Bradbury et al. 2014).
- Lower body mass index (BMI) (e.g. Spencer et al. 2003; Tonstad et al. 2009) and lower risk of obesity.
- Lower chances of developing type 2 diabetes or better management of the disease (e.g. Sabaté and Wien 2010; Trapp and Barnard 2010).
- Reduced risks of developing some cancers (e.g. Key et al. 2014).
- Lower mortality (e.g. Oyebode et al. 2014).

It is the position of the American Academy of Nutrition and Dietetics (2016) that,

> appropriately planned vegetarian, including vegan, diets are healthful, nutritionally adequate, and may provide health benefits for the prevention and treatment of certain diseases. These diets are appropriate for all stages of the life cycle, including pregnancy, lactation, infancy, childhood, adolescence, older adulthood, and for athletes.

## Economic benefits of plant-based diets

In 2015, diets low in fruit and vegetables or high in sugar, processed foods or sodium were estimated to be directly responsible for 37% of all deaths globally and just over a quarter of the total disease burden (disability-adjusted life years (DALYs)) (GBD 2015 Risk Factors Collaborators 2016).

As public health costs are spiralling out of control in multiple countries, plant-based diets have the potential to confer very significant economic benefits. Springmann and colleagues (2016) estimated that 1–31 trillion US dollars, which is equivalent to 0.4%–13% of the global gross domestic product in 2050, could be saved by adopting plant-based diets.

Around 70% of global antibiotics use is applied in farmed animals, who are largely kept in intensive conditions where disease risk is high. This has significantly increased the risk of antimicrobial resistance (AMR). When there are fewer effective antibiotic treatments for humans, the costs to healthcare and risks to human lives could be significant in the not too distant future.

Finally, the growth in plant-based products and plant protein innovation companies is a multibillion dollar business, which makes it an attractive proposition for investors. The vegan protein market is expected to be worth $16.3 billion by the end of 2025, according to market research (Persistence Market Research) from 2017.

## Environmental impacts of the livestock sector

Agriculture covers around 37% of the planet's ice-free land surface (13.4 billion ha) (FAO 2018). Twenty six per cent is used for livestock grazing, and the remaining 11% is used for crop production. Approximately 33% of all croplands is used for livestock feed production.

The production, transport, storage, cooking and wastage of food are substantial contributors to greenhouse gas (GHG) emissions (Intergovernmental Panel on Climate Change 2007; Garnett 2008; Committee on Climate Change 2010). Carbon dioxide is produced from fossil fuels used to power farm machinery and to transport, store and cook foods. The clearing of forests for pasture and feed crop production is also a substantial source. Methane is produced from enteric (intestinal) fermentation within ruminant livestock such as cows and sheep. Nitrous oxide is released from livestock manure and fertiliser. Both methane and nitrous oxide are many times more potent GHGs than carbon dioxide. When measured by consumption (i.e. all GHG emissions related to products consumed in the UK, regardless of where they were produced), food is responsible for approximately one-fifth of all UK GHG emissions (Garnett 2008; Berners-Lee et al. 2012).

Intensive livestock systems may generate fewer GHG emissions per unit of product than extensive systems such as pasture systems, but they have other significant social and environmental impacts, including higher withdrawals of freshwater, more pollution, greater use of antimicrobials with the associated risks of increased antimicrobial resistance, and potentially more outbreaks of zoonotic diseases (FAO 2016a). Also essential in underpinning the modern revolution in food production have been oil- and nitrogen-based fertilisers. Without these, intensive agriculture would not have been possible. However, the environmental impacts of utilising finite resources such as oil are well known and are ultimately unsustainable at present levels (Joy 2017).

The overproduction of food, and of animal production in particular, and the associated environmental impacts have led to severe ecological risks. In 2009, 28 scientists developed the Planetary Boundaries Framework (Stockholm Resilience Centre 2015), which has since been updated. It defines nine planetary boundaries that must not be exceeded in order to protect people and the planet, as

Katherine Richardson has explained in Chapter One of this book. The nitrogen cycle (part of the "biogeochemical flows" boundary) has exceeded the high-risk upper limit (three times the safe limit). Phosphorus is not far behind. Two other boundaries, climate change and land system change, have progressed well into the zone of uncertainty. Animal farming and agriculture are responsible for 70% of freshwater consumption globally, compared to only 22% of water used by industry and 8% for domestic purposes (World Watch Institute 2004).

In fact, the agricultural sector is among the top three global causes of all major environmental problems, including climate change, environmental degradation (pollution, erosion, etc.) and habitat and biodiversity loss (Steinfeld et al. 2006). Due to the inefficiency of converting plant resources into animal-based calories for human consumption, diets rich in animal protein have higher environmental costs.

Increasing human consumption patterns are likely to increase environmental impacts of the livestock sector still further. Tilman and Clark (2014) noted that,

> From 2009 to 2050 global population is projected to increase by 36%. When combined with the projected 32% increase in per capita emissions from income-dependent global dietary shifts, the net effect is an estimated 80% increase in global GHG emissions from food production (from 2.27 to 4.1 Gt per year of $CO_2$-Ceq). This increase of 1.8 Gt per year is equivalent to total 2010 global transportation emissions. In contrast, there would be no net increase in food production emissions if by 2050 the global diet had become the average of the Mediterranean, pescetarian and vegetarian diets.

Similarly, Springmann and colleagues (2016) analysed the health and climate change co-benefits of dietary change to healthier, more plant-based diets on a global level. In line with results from other studies, they found that adopting plant-based (i.e. vegan) diets had the potential to reduce the most GHG emissions (up to 70%). "Healthy global diets" that consisted of lower meat consumption could reduce up to 29% of GHG emissions compared to the FAO reference scenario. When analysing the diets of over 50,000 UK residents, Scarborough et al. (2014) similarly found that dietary GHG emissions of meat eaters were approximately twice as high as those in vegans.

Researchers may use slightly different definitions for diets and apply different methodologies to calculate $CO_2$-equivalent emissions, but the relative difference between diet types is consistent within all studies, with vegan diets providing the greatest environmental benefits.

## Animal agriculture and animal extinction

We are currently living through the sixth mass extinction event since fossil records began (Ceballos et al. 2017). Human activities have increased extinction rates to around 1,000 times that of background levels (Pimm et al. 2014;

Ceballos et al. 2015; Ceballos et al. 2017), and one-fifth of all vertebrate species are now threatened with extinction (Hoffmann et al. 2010).

The multiple causes for this unfolding tragedy are primarily anthropogenic. Land clearing for cities and farms, pollution, overhunting, over-fishing, human overpopulation and climate change have all taken their toll. However, one key factor can be identified which underpins most of these causes: excessive human consumption patterns.

It must surely be considered a tragedy of the highest order, when so many animal species become extinct, never again to walk, fly or swim, above, on or within the Earth or its oceans. Additionally, many of them are important for the maintenance of the ecosystem services – including the clean water, air and healthy environments – upon which all of us depend (Daily 1997).

## Welfare of "food" animals

As mentioned, more than 70 billion terrestrial animals are slaughtered annually (FAO 2017), along with one to three trillion fish (Anon 2014). To meet growing demand, the number of animals farmed for food is expected to substantially increase in the coming decades, with world meat production projected to double by 2050 (FAO 2016b).

Unfortunately, welfare compromises are prevalent within the modern farming of most animal species. Welfare challenges are created by management factors, such as space and environment, and by nutrition; husbandry; access to veterinary care; and limited opportunities to express normal behaviour, including social behaviour. They're also created by animal factors such as genetics and temperament. Many of these factors are exacerbated as farming is intensified to meet growing demand, as described by Joyce D'Silva in chapter 12 of this book.

Welfare problems may also occur when animals are farmed, transported and slaughtered. Finally, most animals farmed for food are killed at a very premature stage of life – foreclosing any future opportunities for achieving positive welfare states or goals that might matter to them.

The space required for feed crop production, or for grazing farmed animals, has also encroached on the natural habitats of wild animals – habitats that are concurrently threatened by pollution, introduced species, hunting and climate change. As mentioned, agriculture now covers around 37% of the planet's ice-free land surface (FAO 2018). As species become endangered or extinct, the individual members of those species may suffer from the effects of habitat destruction and degradation, experiencing hunger, lack of shelter, weakness, disease, increased predation and loss of socially affiliated animals.

## Variance in consumption patterns

Lifestyle choices vary substantially in their ecological footprints. Take grain consumption, for example. Among the most consumptive are the US and Canada, where people consume on average 800 kg of grain annually (most of it indirectly as beef, pork, poultry, milk and eggs). Among the least consumptive is India, where people consume less than 200 kg each (Brown 2009) and, therefore, must ingest nearly all of it directly. Not much of the grain is used for conversion to animal protein, which is an intrinsically inefficient process. As Baroni et al. (2006) put it,

> If animals are considered as "food production machines", these machines turn out to be extremely polluting, to have a very high consumption and to be very inefficient. When vegetables are transformed into animal proteins, most of the proteins and energy contained in the vegetables are wasted; the vegetables consumed as feed are used by the animals for their metabolic processes, as well as to build non-edible tissue like bones, cartilage, offal and faeces.
>
> (Moriconi 2001)
>
> If we only take into account fossil fuel consumption, production of one calorie from beef needs 40 calories of fuel; one calorie from milk needs 14 fuel calories, whereas one calorie from grains can be obtained from 2.2 calories of fossil fuels.
>
> (Pimentel and Pimentel 2003; Reijnders and Soret 2003)

In fact, the Earth already provides enough food for all and could feed at least three billion additional people if the grains fed to animals were used to nourish people directly (Nellemann et al. 2014). Using those grains to produce animal products for wealthier people, whilst others suffer from malnutrition, is a substantial social justice concern.

## Conclusion

Diets high in animal products increase health risks and are responsible for high GHG emissions. In contrast, well-balanced plant-based diets have the potential to substantially save animal and human lives; improve health (e.g. Tilman and Clark 2014), reduce GHG emissions (e.g. Hedenus et al. 2014; Scarborough et al. 2014); and preserve water and land (e.g. Stehfest et al. 2009), and biodiversity.

It may once have been necessary to kill other sentient animals in order to survive. In modern, developed societies, and particularly given the ever-increasing array of animal product alternatives available, this is no longer the case. Vegan diets and lifestyles offer a practical alternative. The Vegan Society (n.d.) defines veganism as "a way of living which seeks to exclude, as far as is possible and practicable, all forms of exploitation of, and cruelty to, animals for food, clothing or any other purpose".

Aspiring towards such a lifestyle will allow virtually all of us to maximise our health and well-being whilst concurrently minimising adverse impacts on the environment and on the other sentient animals with whom we share our planet.

## References


Anon. (2014). *Reducing suffering in fisheries.* http://fishcount.org.uk. Accessed 24 April 2018.

Baroni L, Cenci L, Tettamanti M and Berati M. (2007). Evaluating the environmental impact of various dietary patterns combined with different food production systems. *Eur J Clin Nutr*, 61(2), 279.

Benz-Schwarzburg J and Knight A (2011). Cognitive relatives yet moral strangers? *J Anim Ethics*, 1(1), 9–36.

Berners-Lee M, Hoolohan C, Cammack H and Hewitt C (2012). The relative greenhouse gas impacts of realistic dietary choices. *Energy Policy*, 43, 184–190.

Bradbury KE, Crowe FL, Appleby PN, Schmidt JA, Travis RC and Key TJ. (2014). Serum concentrations of cholesterol, apolipoprotein AI and apolipoprotein B in a total of 1694 meat-eaters, fish-eaters, vegetarians and vegans. *Eur J Clin Nutr*, 68(2), 178.

Brown LR (2009). *Plan B 4.0: Mobilising to Save Civilisation.* New York: WW Norton and Co.

Ceballos G, Ehrlich PR and Dirzo R (2017). Biological annihilation via the ongoing sixth mass extinction signalled by vertebrate population losses and declines. *PNAS*, 114(30), E6089–E6096.

Ceballos G, Ehrlich PR, Barnosky AD, Garcia A, Pringle RM and Palmer TM (2015). Accelerated modern human-induced species losses: entering the sixth mass extinction. *Sci Adv*, 1(5), e1400253–e1400253.

Committee on Climate Change (2010). *The Fourth Carbon Budget. Reducing Emissions through the 2020s.* London: Committee on Climate Change.

Daily G. (Ed.). (1997). *Nature's Services: Societal Dependence on Natural Ecosystems.* Washington, DC: Island Press.

Esselstyn CB (2017). A plant-based diet and coronary artery disease: a mandate for effective therapy. *J Geriatr Cardiol*, 14(5), 317.

FAO (2016a). *The Future of Food and Agriculture: Trends and Challenges.* Rome: FAO.

FAO (2017). *FAOSTAT (Database): Livestock Primary (2017).* www.fao.org/faostat/en/#data/QL. Accessed 03 April 2017.

FAO (2018). *Agricultural Land (% of land area).* https://data.worldbank.org/indicator/AG.LND.AGRI.ZS?view=chart. Accessed 22 February 2018.

FAO. (2016b). *Meat and Meat Products.* www.fao.org/ag/againfo/themes/en/meat/home.html. Accessed 07 April 2018.

Garnett T (2008). *Cooking Up a Storm. Food, Greenhouse Gas Emissions and Our Changing Climate.* Guildford, UK: Food Climate Research Network.

GBD 2015 Risk Factors Collaborators (2016). Global, regional, and national comparative risk assessment of 79 behavioural, environmental and occupational, and metabolic risks or clusters of risks, 1990–2015: a systematic analysis for the global burden of disease study 2015. *Lancet*, 388, 1659–1724.

Hedenus F, Wirsenius S and Johansson DJA (2014). The importance of reduced meat and dairy consumption for meeting stringent climate change targets. *Clim Change*, 124(1–2), 79–91.

Hoffmann M, Hilton-Taylor C, Angulo A, Böhm M, Brooks TM, Butchart SH, Carpenter KE, Chanson J, Collen B, Cox NA and Darwall WR (2010). The impact of conservation on the status of the world's vertebrates. *Science*, 330(6010), 1503–1509.

Intergovernmental Panel on Climate Change (2007). *IPCC Fourth Assessment Report: Climate Change 2007*. Geneva: IPCC.

Joy M (2018). *What is carnism?* www.carnism.org/carnism. Accessed 25 April 2018.

Joy MK (2017). Our deadly nitrogen addiction. In C Massey (Ed.) *The New Zealand Land and Food Annual*. Palmerston North: Massey University Press, p. 2.

Key TJ, Appleby PN, Crowe FL, Bradbury KE, Schmidt JA and Travis RC. (2014). Cancer in British vegetarians: updated analyses of 4998 incident cancers in a cohort of 32,491 meat eaters, 8612 fish eaters, 18,298 vegetarians, and 2246 vegans. *Amer J Clin Nutr*, 100(suppl_1), 378S–385S.

Leenders M, Sluijs I, Ros MM, Boshuizen HC, Siersema PD, Ferrari P, Weikert C, Tjønneland A, Olsen A, Boutron-Ruault MC and Clavel-Chapelon F (2013). Fruit and vegetable consumption and mortality: European prospective investigation into cancer and nutrition. *Am J Epidemiol*, 178, 590–602.

Melina V, Craig W and Levin S. (2016). Position of the Academy of Nutrition and Dietetics: vegetarian diets. *J Acad Nutr Dietetics*, 116(12), 1970–1980.

Moriconi E (2001). *Le fabbriche degli animali: 'mucca pazza' e dintorni [Animal Factories: 'mad cow' and Neighbouring]*. Torino, Cosmopolis.

Nellemann C (Ed.). (2009). *The Environmental Food Crisis–The Environment's Role in Averting Future Food Crises. A UNEP Rapid Response Assessment*. Nairobi: United Nations Environment Programme/Earthprint.

Oyebode O, Gordon-Dseagu V, Walker A and Mindell JS (2014). Fruit and vegetable consumption and all-cause, cancer and CVD mortality: analysis of health survey for England data. *J Epidemiol Community Health*, jech-2013.

Persistence Market Research (2017). *Global Market Study on Plant-Based Proteins: Soy Protein Product Type Segment to Lead in Terms of Market Share during 2017–2025*. www.persistencemarketresearch.com/market-research/plantbased-protein-market.asp. Accessed 15 April 2018.

Pimentel D and Pimentel M (2003). Sustainability of meat-based and plant-based diets and the environment. *Am J Clin Nutr*, 78(Suppl), 660S–663S.

Pimm SL, Jenkins CN, Abell R, Brooks TM, Gittleman JL, Joppa LN, Raven PH, Roberts CM and Sexton JO (2014). The biodiversity of species and their rates of extinction, distribution, and protection. *Science*, 344(6187), 1246752.

Reijnders L and Soret S (2003). Quantification of the environmental impact of different dietary protein choices. *Am J Clin Nutr*, 78(Suppl), 664S–668S.

Sabaté J and Wien M (2010). Vegetarian diets and childhood obesity prevention. *Amer J Clin Nutr*, 91(5), 525S–1529S.

Scarborough P, Appleby PN, Mizdrak A, Briggs AD, Travis RC, Bradbury KE and Key TJ (2014). Dietary greenhouse gas emissions of meat-eaters, fish-eaters, vegetarians and vegans in the UK. *Clim Change*, 125(2), 179–192.

Spencer EA, Appleby PN, Davey GK and Key TJ. (2003). Diet and body mass index in 38000 EPIC-Oxford meat-eaters, fish-eaters, vegetarians and vegans. *Intnl J Obesity*, 27(6), 728.

Springmann M, Godfray HC, Rayner M and Scarborough P (2016). Analysis and valuation of the health and climate change cobenefits of dietary change. *PNAS*, 113(15), 4146–4151.

Stehfest E, Bouwman L, Van Vuuren DP, Den Elzen MG, Eickhout B and Kabat P (2009). Climate benefits of changing diet. *Clim Change*, 95(1), 83–102.

Steinfeld H, Gerber P, Wassenaar T, Castel V, Rosales M and De Haan C (2006). *Livestock's Long Shadow*. Rome: FAO.

Stockholm Resilience Centre (2015). *The Nine Planetary Boundaries*. www.stockholm-resilience.org/research/planetary-boundaries/planetary-boundaries/about-the-research/the-nine-planetary-boundaries.html. Accessed 07 April 2018.

The Vegan Society (n.d.). *Definition of Veganism*. www.vegansociety.com/go-vegan/definition-veganism. Accessed 25 April 2018.

Tilman D and Clark M (2014). Global diets link environmental sustainability and human health. *Nature*, 515(7528), 518–522.

Tonstad S, Butler T, Yan R and Fraser GE (2009). Type of vegetarian diet, body weight, and prevalence of type 2 diabetes. *Diabetes Care*, 32(5), 791–796.

Trapp CB and Barnard ND (2010). Usefulness of vegetarian and vegan diets for treating type 2 diabetes. *Current Diabetes Reports*, 10(2), 152–158.

World Health Organisation (WHO) (2017a). The top ten causes of death. www.who.int/mediacentre/factsheets/fs310/en/. Accessed 22 April 2018.

World Health Organisation (WHO) (2017b). *Obesity and Overweight*. www.who.int/en/news-room/fact-sheets/detail/obesity-and-overweight. Accessed 25 April 2018.

World Watch Institute (WWI) (2004). *State of the World 2004 WWI Report*. Washington DC: WWI.

## Part VIII

# Investing in a flourishing food system

# 32 Corporate engagement strategies to improve farm animal welfare and why they work

*Leah Garces*

In the US, with its 50 ideologically and culturally variable states to negotiate with, legislative progress on farm animal welfare has been incredibly slow at the federal level, though some progress has been made at the state level through ballot initiatives. 2002 saw the first victory to legislate the way farmed animals are confined in Florida. This initiative banned the use of gestation crates. This success set the path for more initiatives in following years. In total 11 state laws have now been passed that ban close confinement of farmed animals in some form. Early initiatives banned one form of confinement but more recent measures have been all encompassing. In 2016 Massachusetts banned the use of battery cages, gestation crates and veal crates as well as the in-state sale of products from any of these confinement systems. What's more, it passed with 77.7% of voters in favour. California is now following Massachusetts' lead and a similar measure is due to be on the 2018 ballot.

While Europe recognised animals, including farmed animals, as sentient beings back in 1997, farmed animals are explicitly omitted from the Animal Welfare Act of the United States (AWA). Under this legislation, they are regulated only when used in biomedical research, testing, teaching and exhibition. The USDA currently interprets the act to exclude birds, all cold-blooded animals (e.g., reptiles), rats and mice bred for research, horses and other farmed animals used or intended for use as food or fibre, like cows and pigs. Apart from those related to consumer safety, there are very few federal level regulations that apply to farmed animals.

Attempts to establish meaningful federal-level regulations in the US have, for the most part, failed. The movement's limited success in this area has included regulation around humane slaughter: in 2002, improvements to the Farm Bill and the Humane Methods of Slaughter Act (originally created in 1958 and amended in 1978) were made, stating that animals must be rendered "fully insensible to pain before they are harvested". This criterion was already written into the original legislation, but was not being adequately enforced. However, even with these updates, the act is widely criticised for excluding chickens, fish, rabbits and other animals routinely slaughtered for food (World Animal Protection, 2014).

In the mid-2000s a new strategy emerged to progress farm animal welfare. Rather than lobbying on an increasingly polarised Capitol Hill, why not engage the food companies directly responsible for the purchase of farmed animal products? Retailers, manufacturers and restaurants – the primary clients of the factory farming industry – became the new focus for advancing the cause. Where federal bureaucracy failed, corporate engagement rose to the challenge and continues to usher in urgently needed change for our food system. Despite the current political turmoil and uncertainties in both the EU and the US, we are currently experiencing a monumental wave of progress for farmed animal welfare through this new approach.

By the end of 2016, over 200 food businesses in the US – including the likes of Walmart, McDonald's and Costco – had committed to switch from caged to cage-free eggs. These pledges are far from trivial: the industry estimates it will cost a whopping $6 billion to build enough cage-free systems to satisfy current demand (USDA, 2016).

By spring of 2018, the same trend was starting to manifest in the chicken industry. Nearly 90 companies and counting have committed to giving chickens raised for meat more space, genetics that don't inherently cause suffering through rapid growth and better living environments – the cost of which would likely dwarf the cost of going cage-free. There are 313 million laying hens in egg production annually, but there are nine billion – that's billion with a 'b' – chickens that move through the meat industry every year. Recognising that billions of animals need better lives would have major financial implications. But despite the obvious risks, these companies have been progressing farmed animal welfare at a pace unheard of in recent history.

To understand why this progress is happening so quickly, we need to perform an MRI on a food company. What makes it tick? What does it value most? The bottom line for companies is, well, the bottom line. The pressure to maintain and increase profit is constant and intense. Unlike an elected official concerned with re-election, a company and its executives are under this pressure quarter to quarter. Most of all, they feel the pressure from consumers, who have seemingly infinite choices as to where to shop and eat. Consumers, should they not agree with the behaviour or purchasing habits of a company, can simply turn to a place that shares their values, in an instant. Not so for an elected official, who often has years before his or her constituents have the chance to vote for an alternative.

Because the marketplace is so dynamic, companies are also under constant pressure from competitors. If a competitor takes advantage of market trends, they soar ahead, while a company that resists meaningful change often falls behind. For the same reason, companies are under pressure from their shareholders and investors: losing patronage to a competitor affects share prices, and that causes investors to balk. Staying ahead of the curve is a constant race, and any misstep or undue delay can be catastrophic.

By tapping into this multifaceted pressure, the dance between staying in touch with public trends and maximising profit at the expense of competitors, we have over and over again seen a domino effect when it comes to companies surging ahead on animal welfare. When one company commits, competitors in that sector – be it food service, restaurants or retailers – quickly follow suit.

In July of 2016, Perdue Farms announced their first animal care policy to work toward many of our goals – enrichments, natural light, more humane slaughter and healthier breeds. And once the company saw the positive response to their progress, animal welfare became a part of company culture. At Compassion in World Farming (Compassion) and WWF's Extinction and Livestock Conference in 2017, CEO Jim Perdue said,

> We are not a small company. The question in the US is: can a big company be trusted? That's a challenge I put to our people every day. I still don't know the answer. If you can get a large company like us [to change], then it gets the attention of our competitors for sure, and will help move the pendulum faster.

That certainly proved true.

In November 2016, Compass Group – the largest food service company in the world – pledged that 100% of their chicken would be fully certified by the Global Animal Partnership by 2024. This means genetics that don't inherently cause suffering through rapid growth, more space, better living conditions and more humane slaughter for all 60 million chickens in their supply chain. Within minutes, Aramark, Compass Group's largest competitor, had done the same. Within months, the entire food service sector had been transformed.

Once the ball was rolling in the food service industry, the restaurant sector took note. Panera Bread, after many months of discussion and a number of farm visits, announced in December 2016 that they, too, would commit to the same improved welfare standards. By spring of the following year, industry giants like Burger King and Subway had also committed. Once food industry leaders had stepped up and a new standard identified, the market had a roadmap and there was no stopping the rest from following it. There was only one direction to go.

Amid this wave of change, Perdue became more and more confident in its position as an industry trailblazer, see timeline in Box 32.1. The company announced in July 2017 that they would provide higher welfare chicken for any client that demanded it. But that wasn't all: other suppliers also started to wake up to the market momentum. Wayne Farms, the sixth largest chicken company in the US, launched a new product line that met Global Animal Partnership standards for chickens. They said they would be there to supply the "Class of 2024" – the tens of companies that had already committed to meet these new animal welfare standards for chickens by 2024. Tyson CEO Tom Hayes, after initially saying that slower-growing birds were not a solution (Meating Place, 2017), did a 180 only a few months later. Perhaps seeing the writing on the wall, he said that if the market demand were high, Tyson would also produce slower-growing chickens (Wattagnet, 2017).

In those few short months, more progress was made for farmed animals than ever before. Niko Pitney of the Huffington Post referred to these improvements as "the most sweeping set of animal welfare measures ever announced" (Huffington Post, 2017). The pace of progress was unbelievable and unexpected.

**Perdue Farms Case Study**

In June 2016 Perdue Farms became the first major US chicken company to publish a detailed animal welfare policy that lays out their current practices and plans for improvement.

This was a surprising and humbling move by Perdue. It came just a year and a half after Compassion released a video of one of Perdue's own contract farmers speaking out against chicken factory farming.

Here's a brief timeline of Compassion's campaign work with Perdue Farms:

DECEMBER 2014 Video of Perdue farmer Craig Watts goes viral. *The New York Times*' Nicholas Kristof covers the story, followed by major media outlets like The Washington Post, HBO's John Oliver and PBS. The story reached over 100 million people (New York Times, 2017).

JUNE 2015 Petition with Craig Watts asking the USDA to stop verifying factory farm chicken as "humanely raised" surpasses 100,000 signatures on Change.org.

AUGUST 2015 Jim Perdue admits, "We need happier chickens". In an acknowledgement that their chickens are unhappy, Jim Perdue tells the NY Times they need to do better (New York Times, 2015).

DECEMBER 2015 Perdue and Compassion begin dialogue to make progress.

Perdue invites the organisation and other advocates to the table to discuss the need for a detailed animal welfare policy (CIWF, 2016).

JUNE 2016 Less than a year after acknowledging they can do better, Perdue commits to:

- Doubling the rate of the birds' physical activity within three years. Importantly, they name slower-growing breeds as one solution.
- Requiring windows for natural light in all new chicken houses and in 200 existing houses.
- Having Controlled Atmosphere Stunning in all Perdue facilities, eventually.
- Sharing transparent metrics to be held accountable for animal welfare.

(Perdue, 2016)

JULY 2017 Perdue releases the company's first annual report on progress since its comprehensive 2016 commitment to accelerate its advancements in animal care. This includes increasing the number of chicken houses with windows, increasing the amount of space given to chickens and implementing a six-hour "lights off" resting period.

(Perdue, 2017)

In essence, animal welfare has been presented as having two possible outcomes for a company: as companies scramble to answer to shareholders, competitors, market trends and consumer demands, animal welfare can either be an enormous risk or an enormous advantage.

Reviewing the shocking footage from inside a caged hen system brings this "risk vs. advantage" dynamic into sharp focus. If a company becomes the subject of an undercover investigation that shows animals trapped in barren cages, feathers missing, unable to turn around – what consumer is going to want to purchase eggs produced by a system so cruel? The question is not "how much is it going to cost to go to cage-free?" but "how much is it going to cost a company to keep the cages?" Essentially, how much market share might be lost by a refusal to change?

Viewing animal welfare through the "risk vs. advantage" lens, we have reached a crucial pivoting point, where it could cost companies more to refuse to do better for farm animals than to continue treating them badly. Consumer awareness and pressure have increased to the point that few are willing to accept animals in cages when presented with the disturbing visuals of what that practice entails. For example, polls from industry trade group Food Marketing Institute (FMI) have shown that animal welfare is critical to today's consumers – and even ranks ahead of environmental concerns. In the 2015 Trends report, FMI reported, "shoppers want food retailers to prioritise animal welfare over environmentally sustainable practices". They advised, "shoppers prioritise animal welfare second only to employment practices" (Food Marketing Institute, 2015).

The numbers don't lie: nobody wants food from animals who have been treated cruelly. More and more companies, along with their shareholders and investors, are recognising that factory farming carries inherent risks for a business's future, including public relations crises spurred by undercover investigations and food recalls or shortages due to disease outbreak (i.e. avian flu).

Thankfully, there is a win-win solution that meets the need to remain competitive in the market, while also contributing to a more sustainable future. As consumers become more concerned about the treatment of the animals in our food system, so must food businesses become attuned to the treatment of animals in their supply chains. Without this critical awareness, a company risks being left in the dust by competitors. And as more companies start to view good animal welfare as a bottom line advantage, the potential to drastically improve the lives of farmed animals becomes enormous. Using a corporate engagement approach that leverages crucial pressure points, it becomes possible that every hen in the world can be let out of her cage, every chicken can enjoy better conditions and healthier genetics that don't lead to a life of suffering. Legislation that puts the nail in the coffin of these archaic and cruel systems then becomes easier, as the market is already there.

Thanks to the inherent competitive forces that drive companies to act, many have risen to the occasion and improved the lives of farmed animals in their care, and a new day for farmed animals has dawned.

An end to factory farming is well within our sight.

## References

Food Marketing Institute, 2015. *US grocery shopper trends*, accessed at: www.fmi.org/docs/default-source/document-share/fmitrends15-exec-summ-06-02-15.pdf

Change.org, 2015. *Petition*, accessed at: www.change.org/p/usda-stop-labeling-factory-farm-chicken-as-humanely-raised

CIWF, 2016. *Perdue first major chicken company to publish detailed animal welfare policy.*

Gabbett. J., 2017. *Tyson CEO not high on 'slow growth' chickens*, accessed at: www.meatingplace.com/Industry/News/Details/73326

Graber, R., 2017. *Tyson foods CEO stresses emphasis on broiler welfare.* WATTAgNet, accessed at: www.wattagnet.com/articles/31568-tyson-foods-ceo-stresses-emphasis-on-broiler-welfare?v=preview The New York Times, Abusing Chickens We Eat, 3 December 2014

The New York Times, *Perdue sharply cuts antibiotic use in chickens and jabs at its rivals*, 31 July 2015.

Perdue, 2016. *Perdue announces industry-first animal care commitments*, accessed at: www.perduefarms.com/news/press-releases/perdue-announces-industry-first-animal-care-commitments/

Perdue, 2017. *Commitments to animal care highlights report*, accessed at: www.perduefarms.com/responsibility/animal-care/2017-highlights-report/

Pitney, N., 2017. *This may be the most sweeping set of animal protections ever announced*, Huffington Post, accessed at: www.huffingtonpost.co.uk/entry/compass-group-aramark-animal-welfare_us_581b6704e4b0b8e11a134130

USDA, 2016. *A momentous change is underway in the egg case*, accessed at: www.usda.gov/media/blog/2016/06/27/momentous-change-underway-egg-case

World Animal Protection, 2014. *Animal Protection Index*, US Country Report, accessed at: https://api.worldanimalprotection.org/country/usa

# 33 The power of investor influence

*Rosie Wardle*

There are many ways to bring about real change and make a difference in the world of farm animal welfare. Indeed, a plethora of these are discussed in other chapters in this book. Campaigning to raise consumer awareness, lobbying governments to implement more stringent welfare legislation and working alongside food businesses to raise the bar of their welfare standards are just some of the vehicles to improve farm animal welfare. But what about driving change using the power of investors?

The financial consequences of factory farming have never been more pertinent than they are today. Large global investors have become increasingly mindful of how companies manage issues such as climate change and human rights.

## The Farm Animal Investment Risk & Return (FAIRR) Initiative

In December 2015, the Farm Animal Investment Risk & Return (FAIRR) Initiative was created to close the knowledge gap that many investors have around the sustainability (or lack thereof) of intensive livestock farming.

Established by private equity entrepreneur Jeremy Coller, the FAIRR Initiative works as a collaborative network for investors to encourage awareness of the financial risks involved in investing in factory farming systems and companies, and to realise the values and rewards associated with other investment opportunities, such as alternative, sustainable proteins.

### *Investors: a vehicle for change*

FAIRR focusses on investors as we recognise their pivotal role as shareholders in the world's largest food producers, retailers and restaurants. Investors can help drive a more sustainable food system by using their influence to address the risks of intensive livestock production and to ensure long-term sustainability.

Today the FAIRR Initiative has over 120 investors participating in its initiatives, collectively managing over $4 trillion of assets. They include large household names such as Aviva Investors, Hermes and Aegon.

By engaging with the companies in their investment portfolios, and highlighting their concern as shareholders, investors encourage meaningful action. This is why FAIRR has organised multi-trillion coalitions of investors to engage with food giants such as McDonalds, Nestlé and Unilever on issues including sustainable protein and antibiotics overuse.

These engagements dovetail with six of the United Nation's Sustainable Development Goals (SDGs):

- Zero Hunger
- Good Health & Well-being
- Responsible Consumption & Production
- Climate Action
- Life Below Water
- Life on Land

FAIRR considers that it is crucial for investors' efforts on sustainability to be aligned with the global development agenda through the SDGs.

The Business & Sustainable Development Commission estimates that $12 trillion in new market value can be unlocked through SDG investing (Business Commission, 2016). Three-quarters of investors are already taking action to support the SDGs, while over 60% think supporting them not only aligns with their fiduciary duty but also offers economic opportunities (UNPRI, 2016).

By aligning their investment and engagement processes with the SDGs, investors are not only looking out for their clients' long-term financial interest but also supporting global development goals. And, as this can also help to improve farm animal welfare, it's a win-win.

## The investment risks

FAIRR helps investors to understand the risks associated with intensive livestock production and to assess these issues as part of their investment processes. A landmark investment risk report (FAIRR, 2016) highlighted 28 environmental, social and governance (ESG) issues associated with intensive farming methods that could be materially significant to investors. These include threatening human health, contributing to climate change and pollution, endangering food security and consuming scarce environmental resources.

## The misuse of antibiotics

The potentially catastrophic threat of antibiotic-resistant bacteria to human health is becoming widely recognised, with the World Health Organisation (WHO) stating, "antimicrobial resistance [AMR] is a global health emergency that will seriously jeopardise progress in modern medicine" (WHO, 2017). AMR is already responsible for approximately 700,000 deaths per year

(O'Neill et al., 2014). Without effective antibiotics, common infections and minor injuries become risky and at times even fatal. It is estimated that antibiotic resistance could cost the world $100 trillion in lost output between now and 2050.

The overuse of antibiotics in the global livestock industry is widely recognised as a factor in this growing issue, and the statistics around this are compelling:

- In the European Union (EU), 70% of antibiotics are used within the animal farming industry.
- In the US this figure exceeds 75% (ECDC et al., 2017).

Globally, the majority of all antibiotics produced today are not given to humans, but to farm animals, often to keep animals healthy in intensive systems in the absence of acceptable farm animal welfare.

Stopping the needless misuse of antibiotics has already led to stricter regulations for livestock producers in both the EU and the US, with further regulation and trade restrictions likely to come. This puts the business models of a wide range of companies across the entire food supply chain at risk.

Starting in 2016, FAIRR gathered 71 institutional investors with over $2.5 trillion in assets under management, to call on global fast food companies such as McDonald's and Wendy's to phase out the routine use of antibiotics in their supply chains. Before the engagement, only one company had a public policy on the use of antibiotics in place. In just over 18 months that figure increased to 16, with UK companies such as The Restaurant Group and Greene King committing to phasing out antibiotics across all types of livestock in their supply chain.

## Meat alternatives: the future of protein production

One of the most important ways in which investors are currently engaging with companies through FAIRR's work is on the issue of the "protein supply chain".

Much of the world relies on meat as a core source of protein. However, with most meat production now coming from factory farming, it is increasingly acknowledged that this system will not meet demand due to its adverse impact on sustainability. From an investment perspective, this could lead to a financial bubble that overvalues farming and food companies and puts trillions of dollars of value at risk.

Such is the extent of the problem that a recent Oxford University study (Oxford Martin School, 2016) calculated that, if unaddressed, public health and environmental expenses associated with increased demand for animal products could cost up to $1.6 trillion globally by 2050.

Consumers too are ever more aware of the negative environmental and health impacts associated with meat production and consumption. As a result, between 1980 and 2012 meat and dairy consumption dropped 15% in the UK (Friends of the Earth, 2014), while 39% of Americans report eating less meat than three years ago (Barclay, 2012).

FAIRR's engagement with food giants such as Nestlé and Kraft Heinz on sustainable protein supply chains – backed by 57 institutional investors managing over $2.3 trillion – aims to highlight the opportunities presented by the growing market for plant-based proteins. The alternative protein market grew by over 8% in 2016 to top $3.1 billion in sales (Byrd, 2017), and demand is set to double by 2024 (Lux, 2016).

Big food companies are taking notice. Pret A Manger, a UK sandwich shop chain, recently trialled a new store selling only vegetarian products and saw sales increase by 70% (Pret A Manger, 2016). Unilever has purchased Sir Kensington's, a sustainable and dairy-free condiment company, while Tyson Foods has acquired a 5% stake in Beyond Meat, a maker of plant-based burgers.

In the UK, following FAIRR's engagement, Sainsbury's has announced that it is trialling new-format stores to encourage customers to choose vegetables over meat, as part of a research collaboration with Oxford University (Weinbren, 2017). Tesco has also recruited a "Director of Plant Based Innovation" to expand their plant-based offerings.

### *Engage, educate, expand*

By engaging with and educating investors about the short-, medium- and long-term risks to their portfolios stemming from the intensive livestock production method, FAIRR's work is already leading to meaningful change within the financial community.

Whether investors are choosing to allocate capital to new, sustainable innovators or, as stakeholders, ensuring that the world's largest food companies are responding to threats to human health and the environment, change is afoot. The long-term losses involved in intensive farming mean it is vital both to the financial community itself and to the wider world that investors continue to step up their responsible investment efforts.

## References

Barclay, E. (2012) 'Why there's less red meat on many American plates', *The Salt*, www.npr.org/sections/thesalt/2012/06/27/155837575/why-theres-less-red-meat-served-on-many-american-plates, accessed 13 December 2017

Byrd, E. (2017) 'New data: Sales climbing for plant based foods', *The Good Food Institute*, www.gfi.org/new-data-sales-climbing-for-plant-based-foods, accessed 13 December 2017

ECDC, EFSA and EMA. (2017) 'Second joint report on the integrated analysis of the consumption of antimicrobial agents and occurrence of antimicrobial resistance in bacteria from humans and food-producing animals', https://ecdc.europa.eu/sites/portal/files/documents/efs2_4872_final.pdf, accessed 13 December 2017

FAIRR. (2016) 'Factory farming: Assessing investment risks,' www.fairr.org/wp-content/uploads/FAIRR_Report_Factory_Farming_Assessing_Investment_Risks.pdf, accessed 13 December 2017

Friends of the Earth. (2014) 'Flexitarianism: The environmentally friendly diet', https://friendsoftheearth.uk/sites/default/files/downloads/flexitarianism-environmentally-friendly-diet-47222.pdf, accessed 13 December 2017

Lux Research. (2016) 'Demand for alternative proteins set to double by 2024', www.luxresearchinc.com/news-and-events/press-releases/read/demand-alternative-proteins-set-double-2024, accessed 13 December 2017

O'Neill, J. and The Wellcome Trust. (2014) 'Review on antimicrobial resistance. Antimicrobial resistance: Tackling a crisis for the health and wealth of nations', https://amr-review.org/sites/default/files/AMR Review Paper—Tackling a crisis for the health and wealth of nations_1.pdf, accessed 13 December 2017

Oxford Martin School. (2016) 'Plant-based diets could save millions of lives and dramatically cut greenhouse gas emissions', www.oxfordmartin.ox.ac.uk/news/201603_Plant_based_diets, accessed 13 December 2017

Pret A Manger. (2016) 'Veggie Pret: What should we do next?', www.pret.co.uk/en-gb/veggie-pret-what-next, accessed 13 December 2017

The Business Commission. (2016) 'Release: Sustainable business can unlock at least US$12 trillion in new market value, and repair economic system', http://businesscommission.org/news/release-sustainable-business-can-unlock-at-least-us-12-trillion-in-new-market-value-and-repair-economic-system, accessed 13 December 2017

UNPRI. (2016) 'What do the UN sustainable development goals mean for investors?', file:///C:/Users/Caroline/Downloads/what-do-the-un-sustainable-development-goals-mean-for-investors.pdf, accessed 13 December 2017

Weinbren, E. (2017) 'Sainsbury's to encourage higher veg consumption with trial', *The Grocer*, www.thegrocer.co.uk/stores/ranging-and-merchandising/sainsburys-to-encourage-higher-veg-consumption-with-trial/547457.article#, accessed 13 December 2017

World Health Organisation. (2017) 'The world is running out of antibiotics, WHO report confirms', www.who.int/mediacentre/news/releases/2017/running-out-antibiotics/en/, accessed 13 December 2017

# 34 The role of business in reducing meat and dairy consumption

*Michael Pellman Rowland*

Over the past years my perch as a contributor at *Forbes* has given me a clear view of the food industry and the developing tides that will shape its future. While there are many game-changing waves reshaping our grocery aisles and fast-food joints, none compare to the tsunami of plant-based products crashing onto the shores of the food world.

## Why is this happening?

There are two primary reasons why we're seeing a flood of animal-free products: sustainability concerns and changing consumer preferences.

Sustainability is a word that can apply to many industries. However, it is most apt when discussing intensive livestock production. From where I stand, factory farming is the most dangerous threat to our species' existence here on earth.

The immense amount of water, land, feed, antibiotics and hormones required to produce meat is painfully unsustainable, as has been documented in other chapters in this book. As emerging economies (namely Asia) build wealth and enter the middle class, their appetite for animal products will surge. This will impose an even greater burden on our fragile ecosystem, which is heating up to dangerously high levels. Look no further than the 17 storms experienced in Houston, Florida and Puerto Rico during 2017 for evidence. Mother Nature is not happy with how we're treating her, and she has every right to be raging mad.

On a positive note, there's light at the end of the tunnel. Informed consumers are increasingly aware of the negative impacts of factory farmed food, and are looking for alternatives. Millennials are by far the most activated consumer group, as the percentage of millennials that are vegan, vegetarian or regularly reducing meat/dairy is skyrocketing (*The Guardian*, 2016). Add to the mix health-conscious consumers, who understand the link between animal protein and a multitude of risk-factors (heart disease, cancer and other diet-related diseases), it's plain to see that the factory farming industry will have a hard time staying the course over the long term. So far this industry has been able to manage risk by exporting to emerging economies, where meat/dairy consumption is increasing whilst scaling down efforts in the developing world where demand is shrinking.

However, with countries like China asking their citizens to reduce meat consumption by 50% (*The Guardian*, 2016), it's likely that the road will get rougher in those markets before long.

## The role that business can play

What excites me most about the challenge of reducing meat and dairy consumption is the role that business can play to accelerate change. We've already witnessed numerous examples of large conglomerates getting on board, charting a course for others to follow.

Whether it was Tyson buying a stake in Beyond Meat, Cargill investing in Memphis Meats or Danone acquiring WhiteWave ($10.4 billion in '16), there has been a swirl of activity around animal-free products.

There's also a number of exciting accelerators like IndieBio and Food Future Co that are helping to pollinate the future food leaders in this space, as are influential investors like Bill Gates and Richard Branson.

Venture capital has also started to make a serious mark in the animal-free sector, allowing companies to expand production, build world-class management teams and enter new markets. We're seeing this at the angel level (New Crop Capital, S2G Ventures) as well as series A & B rounds (Andreessen Horowitz, Khosla Ventures). It won't be long before we see the first 'vegan IPO' ringing the opening bell on Wall Street. Mark my words.

The development of public and private partnerships can really accelerate change. Concordia hosted a summit in New York City in September 2016 on this very topic, in which I participated, specifically addressing the 2018 US Farm Bill and how businesses might work with the public sector to create food policies to actually encourage a system that is better for the planet, consumers and profits. This includes looking at tax incentives for research and development or subsidies that support fruit and vegetables (not just corn and soy that's used to feed animals on factory farms). For example, the fruit industry in America receives 1% of all subsides from the US government. The multiplier effect of business and government working together cannot be overstated.

## Leveraging capitalism

Capitalism has its flaws, but the beauty of this economic system is that companies must adapt or die. To avoid the fates of Eastman Kodak, Blockbuster and the other notable companies that failed to pivot as the world trended past them, companies must stay in step with the consumer. What we're witnessing now is that most of the major food companies are adding plant-based options to their arsenal or expanding the distribution of existing products (some are doing both). And they're doing this for one simple reason: demand.

Data compiled by the Plant Based Food Association and the Good Food Institute (via Nielsen) in 2017 (PR Web, 2017) showcases the impressive growth in animal-free products. What makes the data all the more impressive is that this growth (8.1% growth 2016/2017) occurred while animal-based products declined (0.2% decline during the same period).

The CEO's of food companies and retailers have a simple mandate from shareholders: make money. They want to see sales and profits rise, and they'll do so whether it means selling cow parts or pea protein. In a world of tightening margins and increasing competition, plant-based products are the competitive advantage food leaders must embrace.

## Investors

For the large food companies that have yet to take this seriously, they now have another reason to act: investors. Organisations like FAIRR (see Chapter 33), a group supported by over $4 trillion in assets, are bringing the risk of factory farming to the c-suite. They've started by drawing attention to the threat of antibiotic resistance, something we're likely to hear a lot more about in the coming years. As outlined by Aysha Akhtar (see Chapter 28), we've simply been lucky that we haven't had a health crisis that stemmed from intensive livestock operations. So far, there have been a number of high-profile companies that have succumbed to this pressure and announced plans to phase out antibiotics completely, but more work needs to be done.

Over the years we've heard calls for divestment from alcohol, tobacco, weapons and recently fossil fuels. It was only a matter of time before the sustainability challenges of our food system landed on the radar of institutional investors. After all, they have a fiduciary responsibility to uphold. Holding shares in companies that derive profits from factory farming is akin to holding shares of companies that sold ninja loans leading up the housing crisis of '08. The calls for divestment from companies in the factory farming business are not far off.

## The tide that lifts all boats

The sea change that's occurring in our food system is building, albeit slowly. While the sustainability tide may seem to pose little threat to the large "ships" that dot the horizon, these companies are currently not built for the wave that's coming. The time for action is now.

We are moving faster and faster towards a world that requires us to feed more people with fewer resources. Companies that fail to get onboard this strengthening swell will likely join the long list of failed companies at the bottom of the ocean.

## References

PR Web. *Plant based foods sales experience 8.1 percent growth over past year, September 13, 2017.* Accessed at: www.prweb.com/releases/2017/09/prweb14683840.htm

Milman, O. and Leavenworth, S. 'China's plan to cut eat consumption by 50% cheered by climate campaigners'. *The Guardian*, 20 June 2016. Accessed at: www.theguardian.com/world/2016/jun/20/chinas-meat-consumption-climate-change

# 35 Plant-based and clean meat will save the world

*Bruce Friedrich*

When I first understood that many people in the world don't have enough to eat, I was shocked. I was in eighth or ninth grade, and eating whatever I wanted whenever I wanted was just a given. It had never occurred to me that this wasn't the case for everyone. Ever since then, I have been dedicated to ending global poverty and hunger.

In high school, I thought the solutions were political, and I worked for politicians who I thought would prioritize global hunger relief. In college, I organized fasts to raise money for Oxfam International and volunteered in a homeless shelter and soup kitchen. After college, I spent six years running a homeless shelter in inner city Washington DC and organizing on behalf of global justice.

Now, I run a non-profit organisation focussed on structural change – changing our food system on the supply side in order to allow us to provide more food to more people.

Currently, many things make it difficult to end hunger for our expanding global human family. According to a World Bank report, biofuel production in the early 2000s drove global food prices up 75%, dwarfing the effect of weather changes, drought and other factors (Chakrabortty, 2008). This impact was so severe that Jean Ziegler, the U.N. Special Rapporteur on food policy, called the diversion of crops to biofuels "a crime against humanity" (Ziegler, 2013).

This seems like a fair judgement. Given the hunger and poverty in the world, using crops to make biofuel – and thus depriving people of this food while also driving up the price of the remaining harvest – could certainly be seen as a crime against humanity.

But the extent to which we convert crops into biofuels pales in comparison to the quantity of crops we feed to chickens, pigs, turkeys and other animals. If we are concerned about crops wasted for biofuels, shouldn't we be even more concerned with this much larger diversion of food to animals?

Many people think that feeding global crops to animals is not a big problem, because we then eat the animals. But the vast majority of calories and protein fed to farmed animals doesn't get converted to edible flesh. Instead, the animal uses the crops we feed her to live and grow her entire body – bone, blood and brain – to slaughter weight.

For example, to get one calorie of chicken meat we have to feed the chicken nine calories of crops. That's 800% food waste. It's as if nine plates of food were ready to serve to hungry people, but instead, we threw eight plates away.

Feeding crops to animals is worse than just wasteful. Yes, by eating meat, we lose at least eight out of every nine calories we grow. But by not feeding these crops directly to people, we drive up the cost of the remaining crops even further, exacerbating malnutrition and starvation around the world. It was this impact on the cost of grains – and thus global malnutrition and starvation – that led U.N. Rapporteur Ziegler to call biofuels a crime against humanity. This is also why we should be even more concerned with the practice of feeding crops to animals.

The inherent inefficiency of feeding our food to animals leads to the overuse of other limited resources as well. If we are growing nine times more calories than people are actually able to consume, we are using nine times more land, nine times more water, nine times more fertiliser and nine times more pesticides and herbicides. We're using nine times more fossil fuels to plant, harvest and ship all these extra crops. And we're using more fossil fuels to run factory farms and still more to ship the animals to energy-intensive slaughterhouses.

It may not seem like it when contemplating a crispy chicken sandwich, but animal agriculture is a scourge on our planet. Research by U.N. scientists found that raising animals for food is one of the top contributors to every single one of the most severe environmental problems plaguing us – from water pollution to desertification and deforestation, from biodiversity loss to climate change (FAO, 2006).

The climate impact of our current animal agricultural system is perhaps the most surprising. When we think about global warming, we often think first about our cars. The solution, we're told, is to drive less or buy a hybrid care. But animal agriculture actually causes more climate change than every plane, train and automobile in the world combined (United Nations, 2006). We won't be able to meet our Paris Accord goals without changing our food production. Chatham House – the most widely cited think tank in Europe – declared that governments will be unsuccessful in holding global warming to fewer than 2°C by 2050 unless their populations consume less meat (Wellesley, Froggatt, and Happer, 2015).

Perhaps an even greater danger to human health from animal agriculture is the widespread use of antibiotics on filthy factory farms, which leads to more and more antibiotic-resistant superbugs (Philpott, 2015). These superbugs could cost the world $100 trillion by 2050 (Ellyatt, 2016). A review by the UK government found that the threat to the human race from deadly new drug-resistant disease strains is "more certain" than the threat from climate change (Bingham and Roland, 2014).

The problems of animal agriculture are obvious and overwhelming without considering its direct and immediate impacts on human health or even the animal welfare implications. Tens of millions of Americans – and many more globally – suffer food poisoning from animal products every year. And of course, factory farms and industrial slaughterhouses cause sentient individuals to suffer horribly – so horribly, in fact, that big ag continues to push "ag-gag" laws to criminalise exposing how these animals are treated. They don't want to put laws in place to prevent cruelty to chickens, pigs and turkeys; they want to make it against the law to let the public know about this cruelty.

Given that the current system of meat production is unbelievably wasteful, destructive, poisonous, cruel and deadly, why do we put up with it? Why is animal agriculture growing in almost every single country in the world?

The answer is simple: People want to eat meat.

Luckily, today we have Tofurky deli slices. And Beyond Meat chicken strips. And Field Roast Sausages. And the Impossible Burger. We don't have to poison the planet and ourselves to eat meat. More and more companies are making meat – efficiently and cleanly – from plants. Rather than counting on people to change their diets, these companies are giving consumers the product they want, but a version that is better in every way. These products are now competing directly with conventional meat on the factors that matter to most people: taste, price and convenience.

Here is the story of one of the companies that is re-imagining meat.

Back in 2009, Ethan Brown was working in clean energy because of his concern about climate change. When he learned about the vast harms of industrial animal agriculture, he started asking: What is meat, really? Meat is a combination of lipids, amino acids, minerals and water. Rather than convincing everyone that they should think more about the climate impact of meat, Ethan decided to create climate-friendly meat from plants.

Plant-based meat.

Ethan founded Beyond Meat to test his hypothesis. He hired food scientists, plant biologists, chefs and other culinary experts. He found financial backers interested in his approach. He and his team went beyond the traditional sources of plant-based protein – soy and wheat – and explored many other plants to find just the right sources to bio-mimic animal-based meat.

Three years later, he and his team proved it is possible to make meat from plants.

When Bill Gates – now an investor in Beyond Meat – tried Ethan's chicken strips along with conventional chicken, he was fooled:

> I was surprised to learn that there wasn't an ounce of real chicken it. The 'meat' was made entirely of plants. And yet, I couldn't tell the difference. What I was experiencing was more than a clever meat substitute. It was a taste of the future of food.
>
> (Gates, 2013)

In 2016, Ethan's team created their next masterpiece, the Beyond Burger, which is designed to cook up and taste like a hamburger. The Beyond Burger is actually sold in the meat case at Kroger, Safeway and many other grocery stores. It is also sold in many burger joints. It was restaurant chain TGI Friday's fastest ever test-to-table menu addition.

Virgin founder Richard Branson got on board as an evangelist: "I've had a lot of fun serving juicy hamburgers to everybody at the dinner table and watching their faces as they said, 'This is the best hamburger I've ever eaten,' only to be told it was all vegetarian" (Bercovici, 2017).

Not only has Bill Gates seen the future, but so has the rest of Silicon Valley. Take Eric Schmidt, the former CEO of Google. Recently, Eric was asked to reflect on the top technological innovations he believes will improve life for humanity in the fairly near future. Given his tech background, he talked about the kinds of things we would expect: 3-D printers; self-driving cars; watches that know we're sick before we know we're sick. But the first thing he mentioned – the very first thing – was plant-based meat.

Eric focussed on plant-based meat because of its capacity to feed a growing global population and to do it without the adverse impact on our climate. He calls it "nerds over cattle" (Fehrenbacher, 2016).

Even more telling is that Tyson Foods – by far the biggest meat producer in the U.S. – was so excited about the Beyond Burger that they invested in Beyond Meat. Twice. Former McDonald's CEO Don Thompson has also invested.

Some people think plant-based meat won't ever take over the entire meat market. Perhaps they had a bad encounter with a mushy bean burger back in 1986. They really should try the Beyond Burger or the Impossible Burger, a burger created by former Stanford biochemistry professor Pat Brown (no relation to Ethan). Professor Brown's company, Impossible Foods, has netted investments from the richest person in Asia, Li Ka-shing, Facebook co-founder Dustin Moskovitz, Google Ventures, Bill Gates (again) and many more.

The Impossible Burger is so good that its first public champion was David Chang. About a decade ago, David removed every single vegetarian entrées from all of his restaurants. Now he's featuring the Impossible Burger. If you can win over David Chang with plant-based meat, you can win over anyone.

But for those holdouts who insist that their meat come from animals, other advances in food technology could save us from the myriad harms of animal agriculture. The most promising of these technologies – clean meat – has been a long time coming. Back in 1931, Winston Churchill said, "We shall escape the absurdity of growing a whole chicken in order to eat the breast or wing, by growing these parts separately under a suitable medium".

Churchill's vision is being brought to reality by Mayo Clinic-trained cardiologist Uma Valeti. In 2005, Dr Valeti considered the process he was using to reconstruct heart tissue – taking a few human cells and growing them into healthy tissue for transplantation. Dr Valeti wondered, "Why couldn't we do this with animal flesh and build meat using standard tissue engineering techniques?" In 2015, he launched Memphis Meats around the concept of growing meat directly from animal cells.

This type of meat is called "clean meat" as a nod to clean energy. Clean meat, like clean energy, is produced with far less harm to the environment. Clean meat is also physically cleaner; because there is no slaughterhouse, the meat has no bacterial contamination and no antibiotic or other drug residues.

Almost every month since Memphis Meats launched, the company has driven down the cost of producing clean meat. Bill Gates and Richard Branson are investors, backing both sides of the coming transformation of the meat industry. So is venture capital kingmaker DFJ, an early investor in wildly successful tech companies Skype and Twitter.

Again, the meat industry sees the future. Tyson has invested in Memphis Meats. Major meat conglomerate Cargill, the largest private company in the US, has also invested.

All these individuals and companies are wise to get into the game. Not only are the plant-based and clean meat market sectors going to be incredibly lucrative, they will also save humanity from the threats created by our current system of animal agriculture.

Because of the severity of the threats we face from conventional animal agriculture, governments should also back the development of plant-based and clean meat, given that these technologies are clear solutions to imperative global problems. The U.S. government currently puts $3 billion a year into agricultural research. China puts in even more. Imagine if governments used these resources to form plant-based meat and clean meat research centres at major universities and research centres around the world.

At The Good Food Institute, we're also working to bring about this transformation of the meat industry. We are a non-profit that serves as a think tank, incubator and accelerator for the plant-based and clean meat, egg and dairy market and research sectors. Our team of scientists, entrepreneurs, lawyers and lobbyists recruits more people to these fields while providing start-ups support in everything from branding and marketing to science and policy. Anyone with an idea can come to GFI and get a head start on creating the next world-changing company.

Furthermore, we can each be a part of this, no matter what we personally eat now. This is a trillion-dollar global industry, and the transformation is in its earliest phase. Beyond Meat, Impossible Foods, Memphis Meats and dozens of other companies are growing fast and hiring quickly. They need everyone from MBAs and marketers to scientists and engineers. If you are looking for a good job that can change the world, you need look no further than the plant-based and clean meat future.

There is no reason for anyone to go to bed hungry or worry about their next meal. We can feed the world, but to do so we must replace the current inefficient and destructive means of producing meat. Plant-based and clean meat can give everyone what they want, while improving our health, environment and future.

## References

Bercovici, J. (2017) 'Why Richard Branson and Bill Gates are betting on a food startup you've never heard of', *Inc.*, 26 October [online]. Available at: www.inc.com/jeff-bercovici/memphis-meats-richard-branson.html (Accessed: 4 April 2018)

Bingham, J. and Roland, D. (2014) 'Superbug threat to human race 'more certain' than climate change – inquiry chief', *The Telegraph*, 11 December [online]. Available at: www.telegraph.co.uk/news/politics/11285761/Superbug-threat-to-human-race-more-certain-than-climate-change-inquiry-chief.html (Accessed: 4 April 2018)

Chakrabortty, A. (2008) 'Secret report: biofuel caused food crisis', *The Guardian*, 3 July [online]. Available at: www.theguardian.com/environment/2008/jul/03/biofuels.renewableenergy (Accessed: 4 April 2018)

Ellyatt, H. (2016) ''Superbugs' could cost $1100 trillion – and millions of lives – by 2050: Report', *CNBC*, 19 May [online]. Available at: www.cnbc.com/2016/05/19/superbugs-could-cost-100-trillion--and-millions-of-lives--by-2050-report.html (Accessed: 4 April 2018)

Fehrenbacher, K. (2016) 'The 6 most important tech trends, according to Eric Schmidt', *Fortune*, 2 May [online]. Available at: http://fortune.com/2016/05/02/eric-schmidts-6-tech-trends/ (Accessed: 4 April 2018)

Food and Agriculture Organization of the United Nations (FAO). (2006) 'Livestock's long shadow: environmental issues and options', [online] Available at: www.fao.org/docrep/010/a0701e/a0701e00.HTM (Accessed: 4 April 2018)

Gates, B. (2013) 'Future of Food,' *gatesnotes: the blog of Bill Gates*, 18 March [online]. Available at: www.gatesnotes.com/About-Bill-Gates/Future-of-Food (Accessed: 4 April 2018)

Philpott, T. (2015) 'This Is the Scariest Superbug Yet', *Mother Jones*, 3 December [online]. Available at: www.motherjones.com/food/2015/12/what-uou-need-know-about-superbug-now-breeding-chinas-hog-farms/ (Accessed: 4 April 2018)

United Nations. (2006) 'Rearing cattle produces more greenhouse gases than driving cars, UN report warns', *UN News*, 29 November [online]. Available at https://news.un.org/en/story/2006/11/201222-rearing-cattle-produces-more-greenhouse-gases-driving-cars-un-report-warns#.WgHcxVWnEVM (Accessed: 4 April 2018)

Wellesley, L., Froggatt, A. and Happer, C. (2015) 'Changing climate, changing diet: pathways to lower meat consumption', *Chatham House, The Royal Institute of International Affairs*, 24 November [online]. Available at: www.chathamhouse.org/publication/changing-climate-changing-diets (Accessed: 4 April 2018)

Ziegler, J. (2013) 'Burning food crops to produce biofuels is a crime against humanity', *The Guardian*, 26 November [online]. Available at: www.theguardian.com/global-development/poverty-matters/2013/nov/26/burning-food-crops-biofuels-crime-humanity (Accessed: 4 April 2018)

# Conclusion

*Joyce D'Silva and Carol McKenna*

Now that you have read this book – or at least dipped into it – you too will almost certainly share the concerns which prompted the original Extinction and Livestock conference – and this book itself. You will understand even more deeply the impact of industrial animal farming and its associated industries on the global environment. You will "get" the connection between the massive loss of biodiversity and industrial farming; you will see why iconic species from the Sumatran elephant to the Brazilian jaguar are being squeezed out of their habitats because of our farming methods and our appetites. You will sense the urgency of change.

You will also be encouraged by the stories of down to earth examples of new food and farming approaches in a variety of countries and by the huge moves taking place within the food retail and service sectors. There are definitely signs for cautious optimism.

The truth is we are still struggling to get a global consensus that radical reform is urgent. We don't have a half century to play with. We have encouraging statements from the World Health Organisation (WHO) about dietary change and from the Food and Agriculture Organisation (FAO) and the United Nations Environment Programme (UNEP) about more sustainable and resilient farming methods but, as they say, "the dots" have not really been joined up.

Scientific research shows that the industrial model of agriculture and the Western diet will make it difficult to meet the targets of the Sustainable Development Goals (SDGs), the Paris Climate Agreement and the Convention on Biological Diversity. Without a far-reaching rethink of our food and farming systems it will not be possible to fulfil the objectives of a number of other global agreements including those on rural livelihoods in the poorest countries, food security, the environment, human health, antimicrobial resistance and animal health and welfare. Nor will it be possible to achieve healthy dietary patterns and we will not be able to halt the devastating impact of food production on wildlife.

This is why Compassion in World Farming is committed to working with partners to achieve a global agreement to move to regenerative food and farming systems. This will include Compassion's call that the production and consumption of meat (including fish), milk and eggs from farmed animals should be reduced by 50 per cent in high-consuming nations by 2035 and by half globally by 2050.

We need all the countries in the United Nations to realise the depth of the current crisis and commit to action for a radical reform of our food and farming systems. Of course such global action must take into account national, regional and local differences and starting points. What works in Kenya may not work in Germany etc. But the basic starting points of a wholly sustainable and regenerative farming and food system will meet the same criteria:

An agriculture which:

- Builds healthy soils.
- Uses water efficiently.
- Minimises chemical fertiliser use.
- Minimises pesticide and herbicide use.
- Avoids water pollution.
- Protects local biodiversity and wildlife.
- Minimises GHG emissions.
- Favours mixed farming and agroecological and organic systems.
- Meets high animal welfare standards.
- Minimises the use of antimicrobials in animal farming.
- Favours local and resilient breeds of farm animals.
- Supports the farmers and growers who produce food.
- Improves the livelihoods of small-scale farmers.
- Avoids trade-distorting subsidies and agreements.

The food system adopted should:

- Support food security for all.
- Produce food which contributes to healthy diets and lives.
- Market food in an ethical way, supporting farmers and growers.
- Be priced on True Cost Accounting principles which take account of the adverse impact of industrial farming on the environment, health and animal welfare.
- Empower consumers via transparent labelling as to production method.

Only the UN has the capacity to achieve such a global agreement and to establish monitoring systems to see that it is put into practice.

Compassion in World Farming's CEO, Philip Lymbery, ended the Extinction and Livestock conference in October 2017 by calling for NGOs, academics, businesses, governments and people everywhere to support Compassion's call for a global agreement to transform food and farming systems. We, the editors of this book, repeat that invitation. If you wish to add your voice, then contact: officeofceo@ciwf.org

# Index